Current Trends in Concrete Fracture Research

Current Trends in Concrete Fracture Research

Edited by

Z. P. BAŽANT

Northwestern University
Evanston, Illinois, USA

Reprinted from *International Journal of Fracture*, Vol. 51, Nos 1 and 2 (1991)

Kluwer Academic Publishers

Dordrecht / Boston / London

Library of Congress Cataloging-in-Publication Data

```
Current trends in concrete fracture research / edited by Z.P. Bažant.
      p.    cm.
   ISBN 0-7923-1366-6 (HB : alk. paper)
   1. Concrete--Fracture.  2. Concrete--Cracking.  3. Fracture
mechanics.   I. Bažant, Z. P.
  TA440.C87  1991
  620.1'366--dc20                                    91-802
```

ISBN 0-7923-1366-6

Published by Kluwer Academic Publishers,
P.O. Box 17, 3300 AA Dordrecht, The Netherlands.

Kluwer Academic Publishers incorporates
the publishing programmes of
D. Reidel, Martinus Nijhoff, Dr W. Junk and MTP Press.

Sold and distributed in the U.S.A. and Canada
by Kluwer Academic Publishers,
101 Philip Drive, Norwell, MA 02061, U.S.A.

In all other countries, sold and distributed
by Kluwer Academic Publishers Group,
P.O. Box 322, 3300 AH Dordrecht, The Netherlands.

Printed on acid-free paper

TABLE OF CONTENTS

Foreword

From time to time the *International Journal of Fracture* has presented matters thought to be of special interest to its readers. The last special topic review was presented by Drs W.G. Knauss and A.J. Rosakis as Guest Editors in four issues, January–April 1990, under the general title of *Non-Linear Fracture*. It contained sections on damage mechanisms, interfaces and creep, time dependence, and continuum plasticity insofar as they affect the mechanisms of the fracture process. Continuing this policy, which is consistent with our stated objectives, the two September issues deal with the behavior of concrete and cementious materials during fracture initiation and propagation. We hope that the ensuing state-of-the-art review will yield another instructive and timely product which readers will find useful.

To assist us in presenting this subject, we have prevailed upon a well-known international expert in concrete behavior, Dr. Z.P. Bažant, Walter P. Murphy Professor of Civil Engineering, of Northwestern University to act as Guest Editor. On behalf of the editors and publishers, I wish to thank Professor Bažant and his invited authors for undertaking this special effort.

Pittsburgh, Pennsylvania
September 1991

M.L. WILLIAMS
Editor-in-Chief

International Journal of Fracture **51**: ix–xv, 1991.
Z.P. Bažant (ed.), *Current Trends in Concrete Fracture Research.*

Preface

This preface introduces a special issue of the *International Journal of Fracture* devoted to the recently rapidly advancing field of fracture mechanics of concrete. After various reflections on the current status, twelve research contributions made to this issue by invited eminent experts and their coworkers are briefly summarized and interpreted as part of a broad picture. Concrete fracture mechanics is identified as a maturing field ready for practical applications, which are likely to eventually revolutionize design.

Reflections on current status

Concrete structures were successfully designed and built long before the emergence of fracture mechanics. After that, they have continued to be successfully designed and built without any use of fracture mechanics. But that practice is about to change soon. Recent research has clearly demonstrated that both safety and economy of design will benefit from the use of fracture mechanics. Even more importantly, the use of fracture mechanics will make possible development of concrete materials of higher performance and their utilization in construction.

Concrete is a rather brittle material that cracks under service loads and as a result of regular environmental exposure. In fact, effective utilization of this material requires that extensive cracking in service be permitted. Why then have not concrete engineers made use of fracture mechanics?

Certainly not out of ignorance. Applications of fracture mechanics were attempted beginning in the 1950's, but with unfavorable conclusions. That was of course the linear elastic fracture mechanics and small-scale yielding fracture mechanics, which were developed for metals and are indeed inapplicable to concrete structures except for certain limiting situations such as the behavior at extremely large structure sizes. Consequently, the formulas for brittle failures in the design codes had to remain essentially empirical, based partly on a wrong theory – the plastic limit analysis – and partly on a vast amount of laboratory testing and building experience, which mitigated the inadequacy of the theory. In view of this state of affairs, it has of course been inevitable to impose large safety factors, for which the safety margins were highly nonuniform. Extrapolations of structure sizes, shapes and types beyond the range of previous experience have been unwarranted without extensive testing and development of still further empirical design formulas of limited range.

The difficulty with concrete is that this is not a typical brittle material but a *quasibrittle* material which develops a large zone of distributed cracking that must grow and dissipate energy before a major continuous fracture can form and propagate, as a result of localization of distributed cracking. Fracture mechanics of quasibrittle materials had not started to develop until about 1980. After that, however, the development has been explosive [1–13] and the theory has by now almost matured to practical applicability in design.

The impetus for this recent rapid development came from several directions:

1. The recognition that without some form of fracture mechanics (or equivalent concepts such as nonlocal damage), finite element codes cannot correctly capture damage localization

phenomena and cannot yield objective (mesh-independent) predictions of brittle failures of concrete structures (such as diagonal shear, punching shear, torsion, bar pullout, anchor pullout, fracture of dams, fracture of pipes, cryptodome failure of top slab or reactor vessel, etc.);

2. the realization that these codes cannot correctly predict the effect of structure size on the maximum load, ductility and energy absorption capability in this type of failures;

3. the recognition that the fracture phenomena and brittle aspect are particularly important for modern high-strength concretes or fiber-reinforced cement-based composites and cannot be ignored if the capacity of these new materials should be exploited effectively and safely; and

4. the realization that the way toward developing better cement-based composites cannot bypass understanding of the micromechanisms of fracture, including the spread of microcracks, interface and bond breakages, frictional slip in the fracture process zone, aggregate shear and pullout, pullout of fibers in fiber-reinforced concrete, crack spacing and width, etc.

The fracture problems of concrete are of course not unique to this complex material. Other quasibrittle materials behave similarly – for example modern tough ceramics or rocks. Fracture mechanics of ceramics has recently also been an explosively developing field. Progress in ceramics and concrete can only benefit from mutual interactions. In concrete, the practical experience has a much longer tradition. Concrete and especially the concrete-steel composite are materials that do achieve, in contrast to the low value of tensile strength of concrete, a relatively high fracture toughness and energy absorption capability. To a large extent one can say that the way to achieve these desirable properties with modern ceramics is to emulate concrete, make them behave in fracture more like concrete, e.g. in respect to the role of heterogeneity and weak interfaces, inducement of a large microcracking zone that shields the crack tip, role of volume expansion in the fracture process zone as a toughening mechanism, etc.

In some respects, as a consequence of differences in applications, the fracture theory of concrete probably has been studied more deeply than that of ceramics and advanced farther – for example in the development of nonlocal finite element codes for fracture in structures of arbitrary shapes or in the theory of size effect, which is especially important because of large sizes of many concrete structures. Some transplantations of these results to the field of ceramics might prove profitable. In other respects, though – for example micromechanics analysis and analytical descriptions of various mechanisms of crack-tip shielding or the role of inclusions and their interactions with cracks – the recent research in ceramics became more sophisticated or more advanced. So, vice versa, some emulations in concrete research might prove profitable, although the greater degree of complexity and disorder in the microstructure of concretes will pose limitations.

The bulk of the recent extensive research in concrete fracture has been published in periodicals and conference proceedings [2–13] dealing with concrete. The last International Congress of Fracture in Houston attracted relatively few contributions on concrete fracture, even though a host of specialized conferences devoted to fracture of concrete (and rock) have lately featured over a hundred papers each [1–13]. Obviously, better interdisciplinary communication is needed. Concrete fracture research should be drawn at least partly into the mainstream of fracture mechanics. Therefore, the objective of the present special issue of the *International*

Journal of Fracture is to present representative samples of current research trends in concrete fracture, written by invited eminent experts and their coworkers. The trends may be grouped as follows.

Micromechanics, microscopic observations and rate effect

The effect of a large fracture process zone on the propagation of a tip of a continuous crack can be described basically in two ways:

1. The tip of the actual crack is imagined to be shielded by the surrounding cloud of microcracks, inclusions and other defects, which is described by a decrease in the effective stress intensity factor, or
2. the crack is imagined to extend as a fictitious crack to the end of the fracture process zone but transmit bridging stresses.

The former approach has been prevalent for ceramics, whereas for concrete the latter approach, representing an adaptation and modification of the ideas of Dugdale and Barenblatt, has been preferred. Though physically different, mathematically both approaches give essentially the same results. The latter approach, or an equivalent model for a band of smeared cracking, leads to a simpler fracture model for finite element programs, which explains its popularity in concrete research. Much effort has been devoted to formulation of the softening stress-displacement relation for the crack bridging zone, as well as to microscopic observations and micromechanics solutions that give some clues on the form of this relation.

In the first paper that follows in this issue, F.H. Wittmann and X. Hu discuss measurement and prediction of the fracture process zone using the multi-cutting technique. In a previously loaded fracture specimen, the initial notch is extended by subsequent saw cuts into the fracture process zone, and each time the compliance is measured. From this the stress distribution as well as the softening stress-displacement relation are approximately calculated, after some simplifications. It is shown that the stress-displacement relation is not linear, as assumed in some previous calculations, but has a gradually decreasing slope.

In the second paper, H. Horii and T. Ichinomiya present measurements of the fracture process zone length and the crack opening displacement distribution by the laser speckle technique. They try to separate the role of microcracking in front of the macrocrack tip and crack-bridging behind the macrocrack tip, and conclude that the former is the cause of the pre-peak nonlinearity of the load-deflection curve and the latter is the cause of the post-peak softening. They further argue that the effect of microcracking on the maximum load is less than that of crack bridging and point out that Dugdale-Barenblatt type models do not adequately represent the microcracking zone.

In the third paper, H.W. Reinhardt and J. Weerheijm consider the effect of loading rate on dynamic fracture of concrete. After summarizing the experimental results on the effect of loading rate on the apparent strength, they analyze various influencing mechanisms such as kinetic energy barriers and inertia effects at running cracks. Then they proceed to formulate a model taking into account these effects as well as the flaws in the material, and demonstrate good agreement with test results.

Numerical fracture analysis, design codes and applications

Due to the distributed cracking zones typically observed in reinforced concrete structures prior to failure and even under service conditions, historically the first approach was the smeared cracking model introduced in 1968 by Rashid (cf. [1]). In this model, the material stiffness in the finite element undergoing cracking is reduced either suddenly or gradually after the tensile strength is reached. At first this approach, thought to be more realistic than the analysis of discrete cracks, became very popular and was implemented in large finite element codes. However, after discovering that the results were contaminated by localization phenomena and spurious mesh sensitivity, research interest expanded to fracture modeling by discrete interelement cracks, introduced to finite element analysis of concrete in 1976 by Hillerborg. At the same time, though, it was found that the aforementioned problems of the smeared cracking approach, which is easier to program and more versatile, can be avoided by fixing the element size at the fracture front, as introduced in the crack band model. In 1984 it was shown that a nonlocal approach can achieve the same, with more generality [1]. Thus the smeared crack band and discrete crack approaches became two competing finite element formulations for concrete fracture. Soon, however, it was established that if the displacement across the crack band is matched to the relative opening displacement of the discrete interelement crack, the results of both approaches are equivalent for most practical purposes. This is nevertheless not quite true when fractures do not propagate parallel to mesh lines or when the dominant crack orientation rotates.

This problem is investigated in the fourth paper by J.G. Rots. He considers three variants of the smeared crack band model, in which the cracking direction is fixed, or allowed to rotate with the principal stress direction, or is fixed but multidirectional. The results show that:

1. unlike the rotating smeared cracks, the fixed smeared cracks may lead to unreasonably stiff response; and,
2. unlike the discrete cracks, the smeared cracks in general may cause the so-called stress locking.

In the fifth paper, H.K. Hilsdorf and W. Brameshuber discuss design code formulations that take fracture mechanics into account. This is the most important, final stage of development of concrete fracture mechanics. They focus on the material fracture properties for finite element analysis that need to be specified for design offices, and outline how this can be done to describe the fracture energy dependence on the type of concrete, maximum aggregate size and temperature, and to provide the softening stress-strain or stress-displacement relation for finite element analysis.

The sixth paper, by D.V. Swenson and A.R. Ingraffea, analyzes the collapse of the Schoharie Creek Bridge – well known for its fracture mechanics and size effect aspects. They show that the failure mode must have involved unstable crack propagation in the plinth of the bridge pier, prompted by scour beneath the plinth during a flood. They pay particular attention to the size effect and propose an explanation that differs from the conclusions on the role of size effect reached in a previous study of this disaster.

Nonlinear fracture or damage models and size effect

The seventh paper, by A. Hillerborg, a pioneer in the development of modern fracture mechanics of concrete, reviews the fictitious crack model that he presented with two coworkers in 1976 (cf.

[1]). He discusses its application not only to concrete but also to other materials, including rock and fiber composites.

In the eighth paper, Y.–S. Jenq and S.P. Shah review various features of quasibrittle crack propagation, particularly the energy dissipation mechanisms and their modeling. They also present an extension of their well-known two-parameter fracture model to mixed-mode near-tip fields and loading rate effects. Aside from the requirement of general applicability in finite element codes, they identify the main criteria for the validity of a concrete fracture model to be the notch sensitivity and the size effect.

The ninth paper, by Z.P. Bažant and M.T. Kazemi, deals with the problem of size dependence of concrete fracture energy determined by the work-of-fracture method and adopted as a RILEM recommendation on the basis of a proposal by Hillerborg in 1985. In this method, the fracture energy of the material is determined from the area under the measured load-deflection curve. This curve is calculated on the basis of a master R-curve that is determined, for the given specimen geometry, from the maximum loads of geometrically similar specimens of different sizes (for the post-peak deflections, the actual R-curve is kept horizontal, deviating from the R-curve, as determined in previous research). The calculations show that the fracture energy obtained by this method is indeed strongly size dependent (as well as shape dependent). To get fracture energy as a material constant, the results must be extrapolated to infinite specimen size, in which case the value of the fracture energy must coincide with that obtained by the size effect method (which is now also proposed for RILEM recommendation).

In the tenth paper, J. Planas and M. Elices use softening stress-displacement relations for crack bridging stresses to calculate the size effect. They extrapolate to infinite specimen size by means of their asymptotic method in which the kernel of a singular integral equation is expanded into an asymptotic series, while for small specimen sizes they use a certain influence method based on finite elements. They demonstrate that the size effect depends strongly on the shape of the softening stress-displacement curve for crack bridging.

In the eleventh paper, J. Mazars, G. Pijaudier-Cabot and C. Saouridis make a distinction between two types of size effect – deterministic, which is due to fracture energy release or stress redistributions during stable crack growth prior to the peak load, and statistical, which is due to the randomness of defects in the material. They formulate a continuous damage mechanics model that combines both. The size effect at the initiation of damage is probabilistic, while the size effect in an advanced stage of damage evolution is mainly deterministic and is obtained by means of a nonlocal treatment of damage.

The twelfth paper, by A. Carpinteri, deals with the concept of brittleness number and makes a different distinction between two types of size effect: that on the nominal strength and that on the dissipated energy. A power-law hardening material is considered, and the relation between the effective stress intensity factor and the J-integral is used to deduce the brittleness numbers characterizing the aforementioned two types of size effect and to establish a connection with the R-curve. It is emphasized that in general the structural response depends on both brittleness numbers.

The pervasive theme in the group of the last five papers is the size effect, which no doubt represents the most important consequence of fracture mechanics from the practical viewpoint. Until about 1984 [1] it had been generally believed that any size effect in the observed nominal strength of structures must be statistical, described by Weibull-type theory. This is essentially true of metallic structures, which fail at the initiation of the macrocrack or shortly after that, but not of reinforced concrete structures, which (according to codes) must be designed so that the maximum load is normally much larger than the macrocrack initiation load. This ensures a large stable macrocrack growth prior to the attainment of the maximum load, along with the associated large stress redistributions which inevitably produce a strong size effect. This fact being unknown or unappreciated, the vast majority of testing of brittle failures of concrete structures unfortunately did not include measurements of the size effect, and in those few cases that it did, the size range was insufficient and geometric similarity was not adhered to. (This is a blatant example of the importance of having a proper theory before large testing is undertaken.) Much further testing, therefore, remains to be done in this regard before one can formulate the necessary revisions of the design formulas in codes, so far based on plastic limit analysis which implies no size effect on structure strength.

Concluding remarks

Although the present special issue does not, and cannot, give a complete coverage of all important directions in concrete fracture research, it nevertheless presents a fairly representative picture of the contemporary endeavors in this exciting field. It demonstrates that the discipline has reached maturity, and practical applications lie just around the corner. These are likely to eventually revolutionize the design practice, although many years will likely be required to see it happen.

Acknowledgment

As Editor of this special issue, I wish to express my thanks to all the invited authors for contributing outstanding articles. Preparation of this Preface was supported by NSF Center for Science and Technology of Cement-Based Materials at Northwestern University. Gustavo Gioia, graduate research assistant at Northwestern University, is thanked for valuable assistance in the preparation of this issue.

References

1. ACI Committee 446, *Fracture Mechanics of Concrete: Concepts, Models and Determinations of Material Fracture Properties*, State-of-Art Report, Z.P. Bažant (ed.), American Concrete Institute (ACI), Detroit, Special Publication, in press.
2. Z.P. Bažant, *Applied Mechanics Reviews 39* (5) (ed.) (1986) 675–705.
3. Z.P. Bažant (ed.), *Mechanics of Geomaterials: Rocks, Concrete, Soils*, Proceedings of IUTAM Prager Symposium held at Northwestern University, J. Wiley and Sons, Chichester and New York (1985).
4. L. Elfgren (ed.), *Fracture Mechanics of Concrete Structures*, Report of RILEM Technical Committee 90-FMA, Chapman and Hall, London (1989).
5. M. Izumi (ed.), *Fracture Toughness and Fracture Energy Test Methods for Concrete and Rock* (Preprints, RILEM International Workshop held in Sendai, Japan, Oct. 1988), Tohoku University (1988).
6. V.C. Li and Z.P. Bažant (eds.), *Fracture Mechanics: Applications to Concrete*, Special Publication SP-118, American Concrete Institute, Detroit (1989).
7. J. Mazars and Z.P. Bažant (eds.), *Cracking and Damage* (Proceedings of France–U.S. Workshop held at E.N.S., Cachan, France in Sept. 1988), Elsevier Applied Science, London and New York (1989); for a summary of workshop results see Z.P. Bažant and J. Mazars, *Journal of Engineering Mechanics* ASCE 116 (No. 6) (1990) 1412–1413.

8. H.P. Rossmanith (ed.), *Fracture and Damage of Concrete and Rock*, (Proceedings, International Conference held in Vienna, Austria, July 1988), Pergamon Press, Oxford – New York (1990).
9. S.P. Shah (ed.), *Application of Fracture Mechanics to Cementitious Composites* (Proceedings of NATO Advanced Research Workshop held at Northwestern University, Evanston, Sept. 1984), Martinus Nijhoff, Dordrecht and Boston (1985).
10. S.P. Shah and S.E. Swartz (eds.), *Fracture of Concrete and Rock*, Proceedings of SEM-RILEM International Conference, Houston, June 1987, Springer Verlag, New York (1989).
11. S.P. Shah, S.E. Swartz and B. Barr (eds.), *Fracture of Concrete and Rock: Recent Developments* (Preprints, International Conference held at University of Wales, Cardiff, Sept. 1989), Elsevier Applied Science, London (1989).
12. S.P. Shah (ed.), *Toughening Mechanisms in Quasi-brittle Materials*, Preprints of NATO Advanced Research Workshop, Northwestern University, Evanston, Illinois, July 1990.
13. F.H. Wittmann, (ed.), *Fracture Toughness and Fracture Energy of Concrete*, Elsevier, Amsterdam (1986).

Northwestern University
Evanston, Illinois

ZDENĚK P. BAŽANT
Walter P. Murphy Professor Civil Engineering
Guest Editor

Part 1. Micromechanics, Microscopic Observations and Rate Effect

International Journal of Fracture **51**: 3–18, 1991.
Z.P. Bažant (ed.), Current Trends in Concrete Fracture Research.
© 1991 Kluwer Academic Publishers. Printed in the Netherlands.

Fracture process zone in cementitious materials

FOLKER H. WITTMANN[1] and XIAOZHI HU[1,2]
[1]Institute for Building Materials, Swiss Federal Institute of Technology, ETH-Zurich, CH-8093, Switzerland;
[2]Department of Mechanical Engineering, Sydney University, Sydney, NSW 2006, Australia

Received 1 June 1990; accepted 1 November 1990

Abstract. The current work is directed to the measurement and prediction of the fracture process zone (FPZ) with the compliance method. The method is renovated with a multi-cutting technique. A new compliance curve C_p is established while FPZ in a damaged specimen is removed stepwise by saw-cutting. The length of FPZ can be well determined from C_p in comparison with the compliance calibration curve. Results of two different mortars evidence that the multi-cutting method is applicable to cementitious materials.

A general theory is presented in conjunction with the multi-cutting experiment. The bridging stress transferred within FPZ is evaluated from C_p by the theory. It is proven with a numerical simulation that the strain-softening relation derived from C_p predicts well the global load and displacement relationship. The extension of FPZ at various stages of fracture can also be predicted with the theory if parameters of the strain-softening are available. Both experimental and analytical results affirm the intrinsic connection among FPZ, fracture toughness and fracture energy.

1. Introduction

It is of great interest to fracture mechanics of cementitious and ceramic materials that the fracture process zone (FPZ) defined as the damaged zone ahead of a traction-free crack and the cohesive stress transferred within the FPZ can be measured in experiment. Both the FPZ and cohesive stress have been focuses of fracture research since they are the reasons why the linear-elastic fracture mechanics (LEFM) is invalid for these materials. There is a growing realization that a simple and general method is required with which researchers can measure the characteristics of FPZ not just at best-equipped laboratories, and engineers can easily evaluate the influence of FPZ in a safe assessment of an engineering structure. However, fracture analysis for a structure of a concrete-like material has been far from easy. The difficulty is attributed to several causes. First, there is an intrinsic size dependence in the fracture toughness K_{IC} and the fracture energy G_f. Secondly, many different experimental results, sometimes even contradictory, on these fracture parameters are obtained. To avoid further confusion the role of FPZ in the size dependence of the fracture toughness and the fracture energy has to be fully understood. For this objective both FPZ and the cohesive stress will have to be well determined by experiment.

Based on the assumption that the fracture energy G_f is a material constant, the fictitious crack model (FCM) [1, 2] and the (smeared) crack band theory (CBT) [3, 4] have been developed for concrete-like materials. They represent two extremes in the fracture of unreinforced cementitious materials. FCM considers the fracture in terms of the propagation of a fictitious line-crack. The width of FPZ does not enter into the model. CBT considers the fracture in terms of the propagation of a band of uniformly and continuously distributed cracks with a fixed width. Both FCM and CBT consider the strain-softening behaviour after the tensile strength f_t as the leading root for the presence of the cohesive stress within FPZ. And both have a constant critical crack tip opening displacement w_c. Although FCM and CBT have been successful in numerical simulations of concrete fracture, the size dependence of fracture toughness and fracture energy remains obscure.

To overcome the fundamental problems involved in the application of fracture mechanics to concrete-like materials, the present work is concentrated on the measurement and prediction of FPZ and the cohesive stress transferred within FPZ based on our previous studies [5–8].

2. Experimental details

2.1. Renovation of conventional compliance method

The compliance method is one of the most commonly used techniques and may be the simplest in determination of a crack propagation. However, it is realized that the difference between a saw-cut and a natural crack influences the accuracy of the method considerably. FPZ which commonly exists at a traction-free crack tip in a concrete-like material is the most prevailing factor. In a typical application of the compliance method the influence of FPZ is not separated from that of the traction-free crack. In consequence, the extension of FPZ cannot be determined accurately, and only an effective crack can be determined from a compliance calibration curve. Many attempts have been made to modify the compliance method in search of a solution to separate FPZ from the traction-free crack (for instance [9, 10]). However, the problem remains equivocal.

A simple multi-cutting technique [5, 6] developed from a double-notching method [11, 12] has been introduced into the compliance method, which clarifies the above mentioned uncertainties. Different from the conventional application of the compliance method to determine an effective crack, the essence of the renovated compliance method is to take the saw-cut crack length of a damaged specimen during multi-cutting as the guideline in the comparison of the compliance with the calibration curve. In this way a new compliance curve C_p is established while FPZ is removed stepwise in the damaged specimen. The compliance curve C_p reveals both the length of FPZ and the cohesive stress transferred.

Let us look at a common illustration given in Fig. 1 on FPZ in a concrete-like material. Even for the two dimensional description, the FPZ ahead of the traction-free crack has to be measured in both its length and width. FCM considers well the length influence. The width influence is not clearly shown although the cohesive stress may be viewed as an indirect consideration for the width of FPZ. It is very likely that two different distributions of the cohesive stress will exist in two FPZs if their widths are not the same. To simplify our present analysis, we will concentrate on the length factor of FPZ for the time being. The problem on the width of FPZ will be discussed in a separate paper [13].

As indicated in Fig. 1, a damaged material can be asumed as a component constituted of materials of two different Young's moduli. Let E be the Young's modulus of the undamaged material and E^* be the effective Young's modulus of the damaged region. If FPZ has been developed in one of two identical specimens with the same saw-cut crack, a difference in compliance will be seen due to the Young's modulus diminution as indicated in Fig. 2 where P is the external load and $CMOD$ is the crack mouth opening displacement. Figure 3 illustrates how the length of FPZ can be determined with the compliance method in

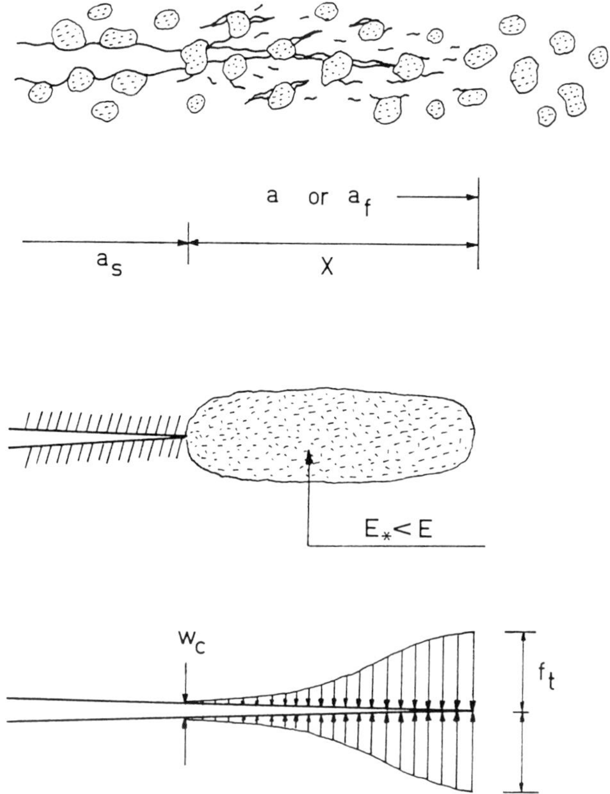

Fig. 1. Simplified two-dimensional representation of the fracture process zone (FPZ) in a concrete-like material; the effective Young's modulus E^* within FPZ is reduced due to cracking; bridging stress diminishes towards the crack tip.

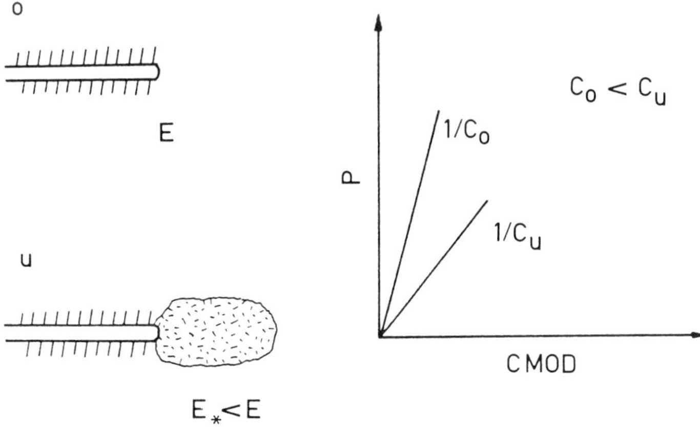

Fig. 2. Schematic representation of *P-CMOD* curves for an undamaged and a damaged specimen. The compliance of the damaged specimen is reduced due to the softening zone in FPZ.

conjuction with multi-cutting, where C indicates the compliance calibration curve, and C_p denotes the compliance of a damaged specimen obtained from the multi-cutting process. The x-axis in Fig. 3 indicates the increment of the saw-cut crack with the origin at the initial saw-cut crack tip. With the stepwise cutting, FPZ and its influence can be well

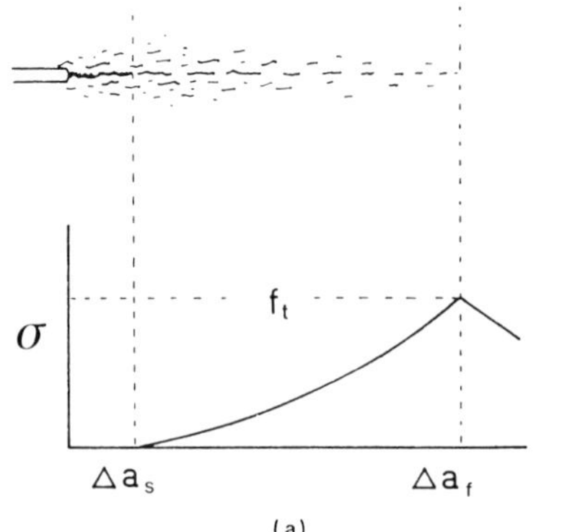
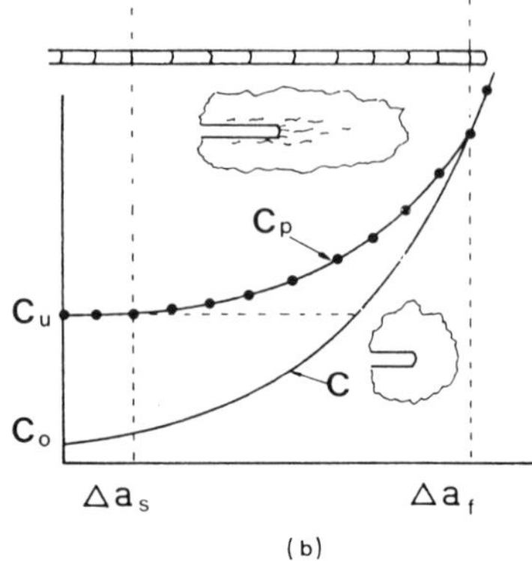

Fig. 3. Determination of FPZ with modified compliance method: (a) the length of FPZ and bridging stress distribution; (b) stepwise cutting into FPZ and the resulting compliance.

separated from the traction-free crack. Before any bridging stress is removed, the compliance value C_u at an unloading point will remain constant. The compliance will be increased if any bridging stress is removed by cutting. If the cutting is through the whole softening region, the compliance will be on the calibration curve. Let Δa_s denote the propagation of the traction-free crack, and Δa_f the total crack propagation including the extension of FPZ. The length of FPZ is simply given by $\Delta a_f - \Delta a_s$.

2.2. Experimental verification of fictitious crack

The multi-cutting method outlined in the previous section is used to determine the extension of FPZ in wedge opening loaded or wedge splitting test [14] specimens of mortar. The dimensions of specimens are 200/197/15 mm. W, the distance from the loading point to the back surface of the specimen along the saw-cut crack direction, is 183 mm. To control the crack propagation along the anticipated saw-cut crack direction, the specimens are grooved 2 mm in depth on both sides.

Two different mortars are used in the experiment. Mortar A has a water/cement ratio of 0.4 and a sand/cement ratio of 1.5. The maximum diameter of sand is 4 mm. Specimens have been cured in water for 48 days before testing. Mortar B has a water/cement ratio of 0.5 and a sand/cement ratio of 1.5. The maximum diameter of sand is 1 mm. The curing time for mortar B is 69-days.

Experiments are performed immediately after the specimens are removed from water. Results of both mortar A and B are shown in Fig. 4. The change of compliance attributed to the removal of the bridging stress is clearly demonstrated in both materials. The initial a/W ratio selected for the specimen of mortar A is 0.4 (with an initial ligament of 110 mm), and 0.5 for the specimen of mortar B (with an initial ligament of 91.5 mm). The results are presented in terms of the length of recutting with the origin at the initial saw-cut crack tip of the specimens. It is obvious that special care should be taken to ensure that no further extension of FPZ is induced during cutting and reloading; for instance, a fast cutting speed and a low feeding rate. And the reloading should always be less than the critical level at which a new extension of FPZ will occur. The linear portions of both loading and

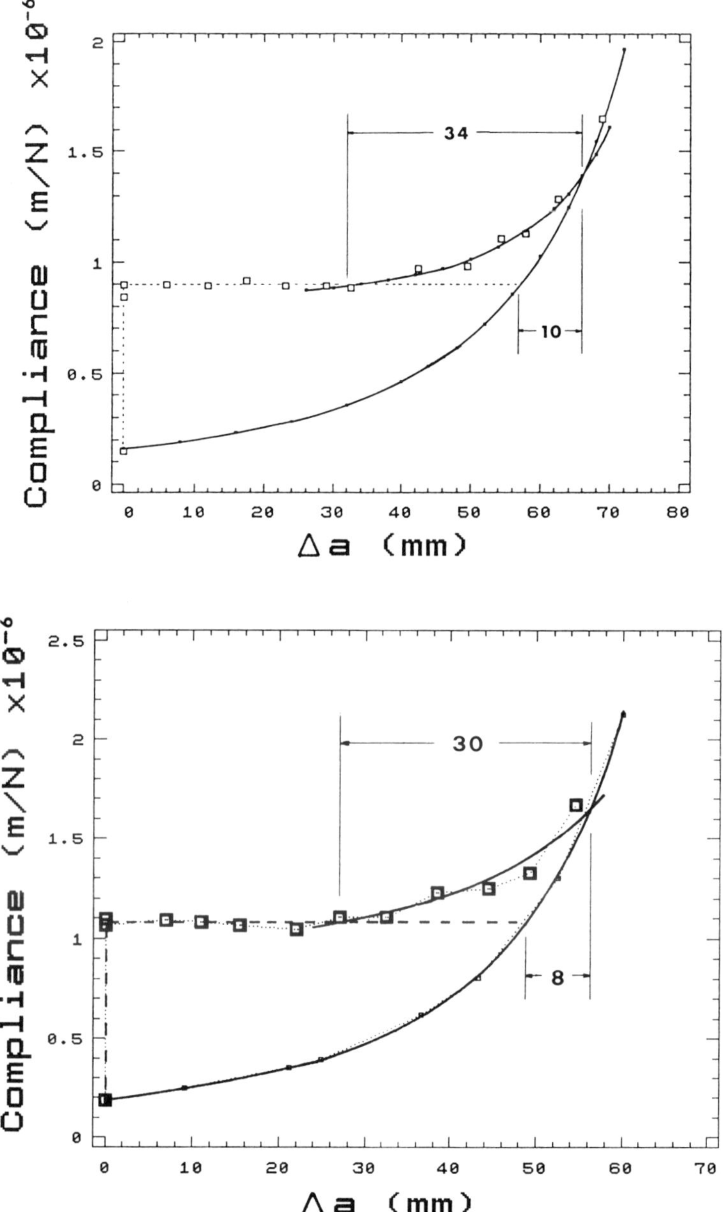

Fig. 4. FPZ determined with modified compliance method: (a) mortar A specimen with ligament of 110 mm; (b) mortar B with ligament of 91.5 mm.

unloading branches of a *P-CMOD* curve are used to monitor the change of the compliance C_p. Details on the compliance evaluation are given in [5, 15].

Results of both mortar A and B in Fig. 4 give evidence that FCM is adequate in modelling of FPZ in terms of the softening behaviour after f_t, and the renovated compliance method is generally applicable to concrete-like materials. Indeed, both Δa_f and Δa_s defined in Fig. 3 can be well determined from the multi-cutting experiment. The remaining task, however, is to find out the cohesive stress distribution between Δa_f and Δa_s.

3. Fracture process zone modelling

3.1. Conditioned FCM

FCM is based on assumptions that the fracture energy G_f is a material constant and the strain-softening $(\sigma - w)$ curve is unique. Under these assumptions a $\sigma - w$ curve determined in a direct tensile test is unconditionally applied to material under other loading conditions. It should be pointed out that G_f, according to its definition [16], is an averaged energy quantity over a projected fracture area. G_f is only meaningful as a description of a local energy consumption if the total energy in fracture is more or less uniformly distributed over the area. This definition of G_f appears unsophisticated in some circumstances. The size of FPZ (both its length and width) is directly related to G_f since the damaged region is where the fracture energy is consumed. For most of the specimens used nowadays in the determination of fracture parameters of cementitious materials, the size of FPZ is not small compared to the dimensions of the specimens. Consequently, a change in the size of FPZ will not be negligible. It is speculative that the change in the size of FPZ will lead to a non-uniform distribution of the total fracture energy throughout a broken ligament. Therefore, even the ligament can presumably be divided into thin strips under tension; they are not going to go through the same stress condition as they are supposed to do in a direct tensile test. The damaged zones in these presumed strips are different due to the variation in the width of FPZ. The problems on G_f and the width of FPZ will be discussed comprehensively in [13].

The relationship between the tensile strength f_t and micro-fracture presumed in Figs. 1 and 3 is another concern. f_t as the maximum global tensile strength is always higher than the micro-tensile strength at which non-local stable micro-fracture occurs for two-phase materials like concrete and mortar [17, 18]. It has been shown by an interferometry observation in direct tension of concrete [19] that a localized cracking later developed into FPZ occurs at about 90 percent of f_t. If time-dependent fracture is not considered and if G_f is the major concern, it is acceptable that f_t is about the same as the micro-tensile strength. In other words, the strain hardening before f_t can be neglected in FCM as an appropriate approximation since the energy dissipation before f_t is considerably smaller than the fracture energy G_f when a time dependent fracture is excluded.

Under certain conditions the total fracture energy will be relatively uniform throughout the ligament and the width of FPZ will vary little. A thin thickness and a deep groove, such as the specimen dimensions adopted in the present study, help to achieve a constant G_f and limit the variation in the width of a FPZ. Therefore, G_f in this paper is taken as a material constant, and no variation in the width of FPZ is taken into account.

3.2. Cohesive stress transferred within FPZ

For a concrete-like material the assumption of a linear crack profile is adequate since the displacement of the crack surface is relatively small. The applicability of the assumption has been proven both experimentally and numerically [20, 21].

Following the definition of C_p in Section 2.1, a general C_p curve is shown in Fig. 5 together with the calibration curve C. If no bridging stress is removed, C_p is constant. Both C and C_p will merge together after the whole bridging stress within the strain-softening zone is removed.

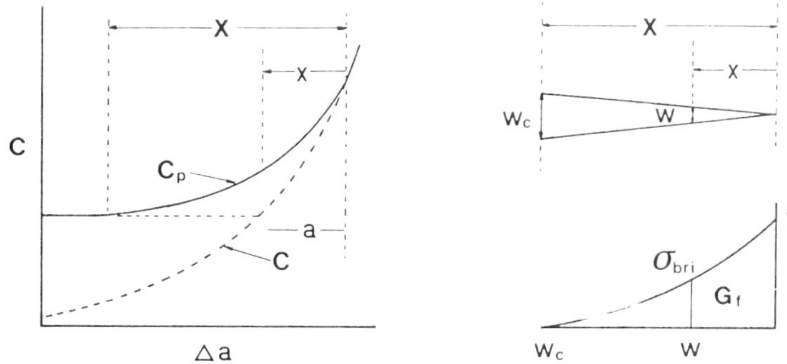

Fig. 5. Relation between compliance C_p from a damaged specimen and calibration curve C, and bridging stress transferred within FPZ.

Let X be the length of FPZ and w_c be the critical crack tip opening displacement. From the linear crack surface model, we have

$$\frac{x}{X} = \frac{w}{w_c}.\tag{1}$$

Then the fracture energy G_f illustrated in Fig. 5 is given by

$$G_f = \int_0^{w_c} \sigma_{\mathrm{bri}}\, dw,$$

$$= \frac{w_c}{X}\int_0^{X} \sigma_{\mathrm{bri}}\, dx.\tag{2}$$

It is shown in Fig. 6 that the crack opening profile will not be influenced by the removal of the bridging stress during cutting if $CMOD_u$ at the unloading point is kept constant. In this case

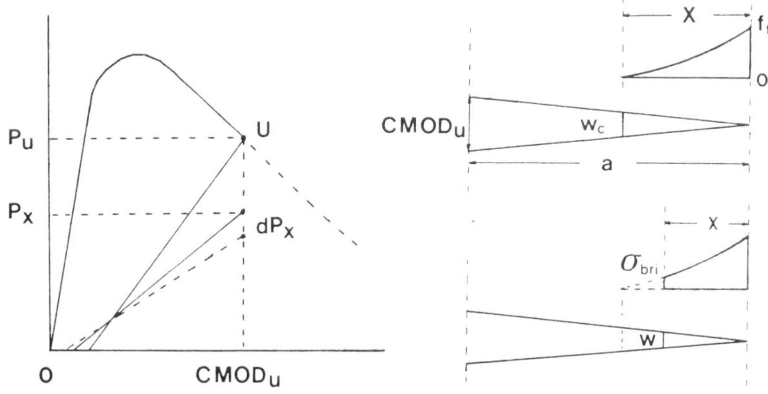

Fig. 6. Relation between load-displacement curve and the removal of bridging stress during consecutive cutting.

the external load P has to be reduced to compensate for the removal of the bridging stress. Under this condition the bridging stress σ_{bri} will remain as it is at the unloading point. The relation of $\sigma_{bri} - w$ (or $\sigma - w$ of the strain-softening) can then be related to the change of the compliance determined in terms of P_x and $CMOD_u$ as illustrated in Fig. 6. Thus it will be an advantage that for the following analysis the crack mouth opening displacement is kept constant as $CMOD_u$.

Let us consider the P_x-$CMOD_u$ relationship after part of the FPZ has been removed by cutting. For the remaining $CMOD$ when the external load $P = 0$ is neglected (later it can be seen that $CMOD_u -$ constant will give the same result), the relationship is simply given by

$$C_p(x) = \frac{CMOD_u}{P_x}, \tag{3}$$

or

$$dC_p(x) = \frac{CMOD_u}{P_x^2}(-dP_x) = \frac{C_p^2(x)}{CMOD_u}(-dP_x). \tag{4}$$

Here it is clear that the assumption of keeping $CMOD$ constant as $CMOD_u$ also helps to simplify (4). Since $dC_p > 0$ during cutting, $dP_x < 0$ as implied in Fig. 6. From curve C_p in Fig. 5, we have

$$dC_p(x) = \frac{dC_p}{da}da = C_p'da. \tag{5}$$

Here da is the saw-cut length corresponding to the change of $dC_p(x)$. dC_p/da rather than dC_p/dx is used since the calibration curve C normally is constructed for either Δa or a/w. But it can be easily obtained from Fig. 5 that $dC_p/da = -dC_p/dx$ since $da = -dx$. It can be seen from (4) and (5) that if the FPZ is removed by the amount of da, P_x must be reduced by dP_x to keep $CMOD$ constant as $CMOD_u$. The relationship between dP_x and the removed bridging stress is not simple since the stress in the linear elastic part of the material will be redistributed due to the change of the neutral axis. Hence we introduce a coefficient k so that

$$-dP_x = \frac{1}{k}\sigma_{bri}Bda. \tag{6}$$

Here B is the thickness of the material. From (4), (5) and (6) we have

$$\sigma_{bri} = k\frac{CMOD_u}{B}\frac{C_p'}{C_p^2(x)}. \tag{7}$$

In general, k is a variable. The total crack length a and the position x within FPZ are needed to describe the distance between the external load P and the bridging stress

$\sigma_{bri}(x)$. Therefore, it can be written that $k = k(a, x)$. Equation (2) shows that the integration of the bridging stress over the FPZ is equal to the fracture energy times X/w_c. Substituting (7) into (2) we have

$$G_f = \frac{w_c}{X} \int_0^X k(a, x) \frac{CMOD_u}{B} \frac{C_p'}{C_p^2} \, dx,$$

$$= \frac{w_c}{X} \frac{CMOD_u}{B} \int_0^X k(a, x) \frac{-dC_p}{C_p^2},$$

$$= \frac{w_c}{X} \frac{CMOD_u}{B} \left\{ \frac{k(a, x)}{C_p(x)} \Big|_0^X - \int_0^X \frac{k'(a, x)}{C_p(x)} \, dx \right\}. \tag{8}$$

Here the relationship $dC_p/da = -dC_p/dx$ is used. Equation (8) can be simplified for two occasions. First, the length of the FPZ X is much smaller than the crack length a. In this case $k(a, x) = k(a) = $ constant for a given crack size a. Second, dk/dx is small so that the second part of integration in (8) can be neglected. As a first approximation the relationship $k(a, x) = k(a)$ can be used. For a given crack size a, $k(a)$ is a constant and is independent of x. $k(a)$ can then be averaged in terms of the fracture energy, and is given by

$$k = \frac{G_f X B}{w_c CMOD_u} \frac{C_p(0)C_p(X)}{[C_p(0) - C_p(X)]}. \tag{9}$$

It should be noticed that the simplified expression of $k(a, x)$ given in (9) still depends on the crack size a. The length of the FPZ X, $C_p(0)$ and $C_p(X)$ together show the influence of the crack size a. If the $k(a)$ given by (9) does not vary very much over a large range of Δa ($> x$), $k(a, x) = k(a)$ will be justified. From (7) and (9), the bridging stress is given by

$$\sigma_{bri}(x) = G_f \cdot \frac{X}{w_c} \cdot \frac{C_p(0)C_p(X)}{[C_p(0) - C_p(X)]} \cdot \frac{C_p'(x)}{C_p^2(x)}. \tag{10}$$

The significance of (10) is that it links the local bridging stress, the fracture energy G_f, w_c and the length of FPZ together with the compliance C_p, the global behaviour of the specimen.

Since a is the total crack length which includes the FPZ, it is obtained from Fig. 5 that

$$C(a) = C_p(0),$$
$$C'(a) = C_p'(0). \tag{11}$$

That is $x = 0$ at the tip of the total crack length a. From Figs. 5 and 6 it is clear that the tensile strength f_t of the material will be given by (10) if $x = 0$. Thus we have

$$f_t = G_f \cdot \frac{X}{w_c} \cdot \frac{C_p(X)}{C_p(0)} \cdot \frac{C'(a)}{[C_p(0) - C_p(X)]}. \tag{12}$$

Here we use $dC(a)/da$ instead of $dC_p(0)/da$ because they are equal, and the error in C_p can be minimized by using C' at the same point. From (10) and (12) it can be obtained that

$$\frac{\sigma_{\mathrm{bri}}(x)}{f_t} = \frac{C_p^2(0)C_p'(x)}{C_p^2(x)C_p'(0)},$$

$$= \frac{C^2(a)C_p'(x)}{C_p^2(x)C'(a)}, \quad 0 < x < X. \tag{13}$$

Equation (13) shows that the ratio of the bridging stress over the tensile strength of the material only depends on the compliance curve C_p. Since the tensile strength f_t is a material constant, the bridging stress within the FPZ can be determined solely with the compliance curve C_p. This result is anticipated as we know C_p is obtained through removing the bridging stress within the FPZ. Thus after construction of the compliance curve, C_p, both the length of the FPZ X and the ratio of the bridging stress over the tensile strength can be well determined. To determine the tensile strength f_t the fracture energy G_f needs to be measured separately. w_c can be determined from Fig. 6 with the knowledge of both $CMOD_u$ and X. As an alternative, the ratio of G_f/w_c can be determined by (12) if the tensile strength f_t is determined in a direct tensile test (stable fracture after f_t is not required). It is clear that (12) and (13) link FPZ and the strain-softening, which makes FPZ analysis much easier than before.

Since there is no $CMOD_u$ in (10), the remaining $CMOD$ when $P = 0$ will not affect our last result. Even $CMOD_u$ in (3) is replaced by '$CMOD_u$ − constant'; later it will be cancelled. When $x = X$, $C_p' = 0$; (10) shows that the bridging stress is zero as expected. It should be mentioned that (13) is valid even if the unloading point U is before the P_{max} in Fig. 6. If a FPZ is not fully

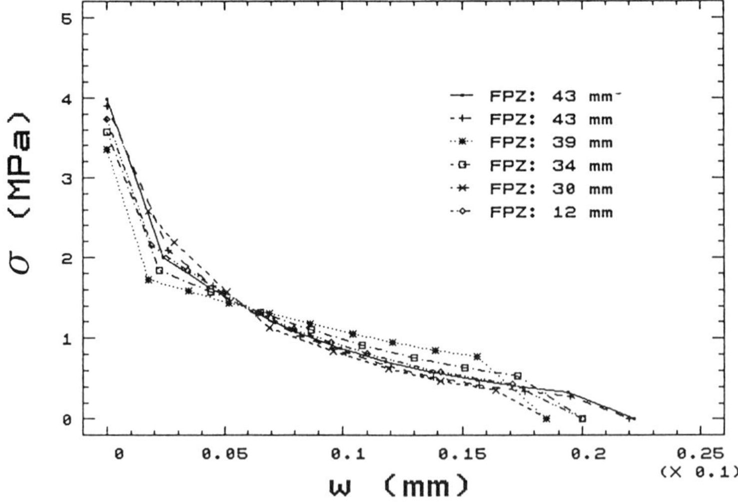

Fig. 7. Bridging stress (or strain softening) determined from compliance C_p curves of mortar A specimens; results from six individual tests with different extension of FPZ are shown.

developed, $C'_p > 0$ even at the initial saw-cut crack tip. In this case the length of the FPZ can also be estimated by comparison of the bridging stress obtained at the initial saw-cut crack tip with the value of the strain softening curve.

Results of the bridging stress distribution for mortar A determined from (13) are shown in Fig. 7. The measured G_f is 21.8 N/m and the average w_c is 0.02 mm. The average f_t calculated from (12) is 3.71 MPa, which is commonly observed for mortar and concrete. Although the length of FPZ varies from 12 mm to 43 mm, the same $\sigma - w$ curve is found. Details can be found in [8, 15]. The results in Fig. 7 imply that the assumption of a constant G_f is adequate in this investigation. It is clear with those results that the strain-softening relationship can truly be determined by C_p. Equation (13) has been further proven by a numerical simulation [22]. In the simulation not only the multi-cutting experiment has been successfully run by computer, *P-CMOD* curves in good agreement with experiment have also been reproduced with the strain softening relation derived from (13). It is worth pointing out that C_p curves simulated numerically [22] are very similar to those determined in experiment. The agreement affirms the argument on the compliance evaluation [5, 15].

4. Prediction of FPZ with compliance method

4.1. Fully developed FPZ

If a FPZ has been fully developed as illustrated in Figs. 3 and 5 then $\Delta a_s > 0$, $C_u = C_p(X)$. Then (12) can be used to predict the extension of FPZ. From the linear crack surface assumption, it is obtained that

$$X = \frac{w_c}{CMOD_u} a. \tag{14}$$

From (12) and (14) we have

$$a = CMOD_u \frac{f_t}{G_f} \frac{C(a)}{C'(a)} \left\{ \frac{C(a)}{C_u} - 1 \right\} \tag{15}$$

noting $C(a) = C_p(0)$. For given C_u and $CMOD_u$ in a *P-CMOD* curve, the total crack length a in (15) can be determined through iteration. The corresponding FPZ X can then be obtained with (14).

If should be pointed out that the entire $\sigma - w$ curve is not prerequisite in the prediction of a fully developed FPZ. Only the ratio of G_f/w_c and f_t is required. This advantage is attributed to the generality of the experimental and analytical methods.

4.2. FPZ under development

Before a FPZ has been fully developed ($\Delta a_s = 0$ in Fig. 3), an extra equation is required to determine the extension of FPZ. The reason is that $C_p(X)$ remains as an unknown in (12).

However, the problem can be solved by assuming a strain-softening relationship. Let a_0 be the initial crack length, and X_u be the length of FPZ at an unloading point.

$$a - a_0 = X_u < X. \tag{16}$$

Here X should not be confused with the length of FPZ X_u. It is only a distance indication from the tip of the total crack including FPZ, where the crack surface separation is w_c. It has been shown [7] that

$$\frac{C(a)}{C_u} = 1 + \frac{w_c}{CMOD_u} \frac{a}{[n+1]} \frac{C'(a)}{C(a)} \left\{ 1 - \left[1 - \frac{CMOD_u}{w_c} \frac{(a-a_0)}{a} \right]^{n+1} \right\}, \tag{17}$$

where

$$n = \frac{f_t w_c}{G_f} - 1 \tag{18}$$

obtained in terms of an empirical power law strain-softening [21]. The exponential n in (18) is a parameter averaged from w_c/G_f and f_t even if the true strain-softening is not a power-law. For given C_u and $CMOD_u$, the total crack length a can be solved. The length of FPZ X_u can then be obtained with (16).

5. Results and discussion

One of *P-CMOD* curves of mortar A is shown in Fig. 8. The initial a/w ratio of the specimen is 0.4 (with a ligament of 110 mm). As usual, a K_r curve given in Fig. 9 is constructed for the effective crack propagation which is determined by the compliance calibration curve in a normal way. It can be seen that K_r first increases to its plateau, then remains constant, but finally fails. Predicted lengths of FPZ from (15) and (17) for numbered points in Figs. 8 and 9 are shown in Fig. 10 together with measurements from several other specimens [8]. Evaluated results of FPZ

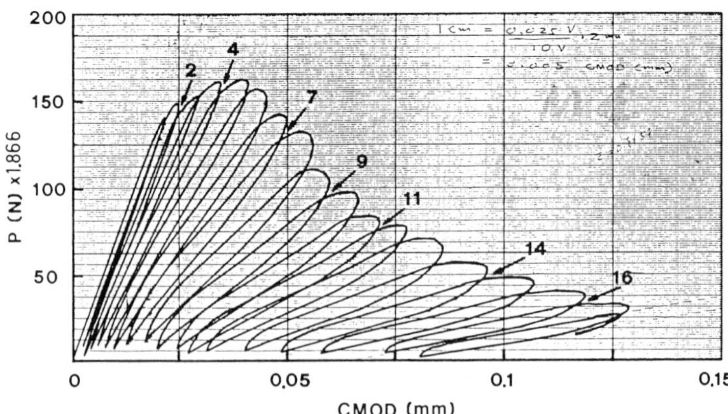

Fig. 8. Typical load-displacement curve of mortar A specimen with initial a/W ratio of 0.4 (with ligament length of 110 mm).

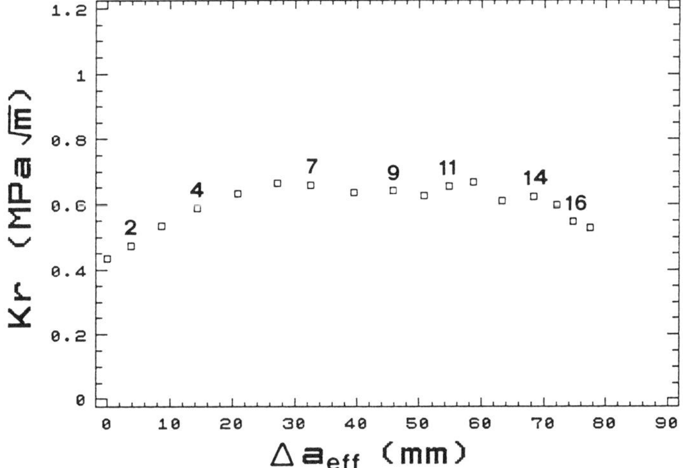

Fig. 9. K_r-curve constructed with effective cracks determined from unloading loops in Fig. 8 by compliance calibration curve.

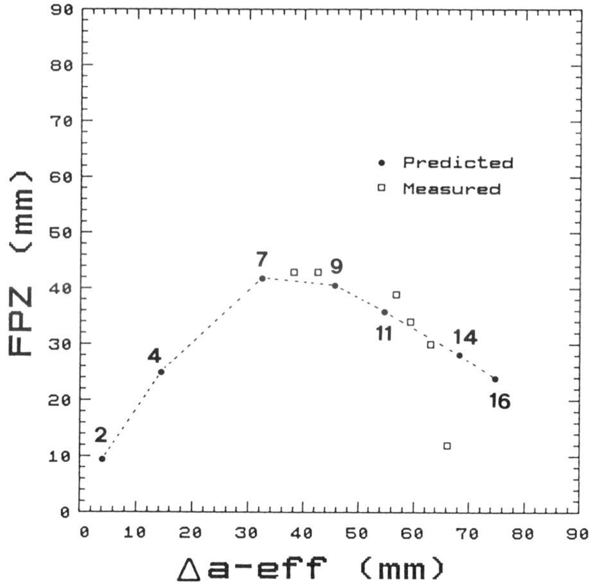

Fig. 10. Comparison of predicted lengths of FPZ for numbered unloading loops in Fig. 8, and measured lengths of FPZ in several specimens with modified compliance method.

show that the extension of FPZ is behind the rising and falling K_r-curve. Holographic interferometry has been used to measure the extension of FPZ in a similar mortar specimen [23] (see Fig. 11). The result of around 23 mm at a maximum load is very close to the corresponding FPZ of 25 mm (point 4 in Figs. 8 and 9) given in Fig. 10.

From Fig. 10 it is clear that the length of FPZ is a variable even if the fracture energy G_f is assumed to be constant. The extension of FPZ is bound to be influenced by the ligament. In consequence, a falling K_r-curve is observed. And the fracture toughness is ligament dependent

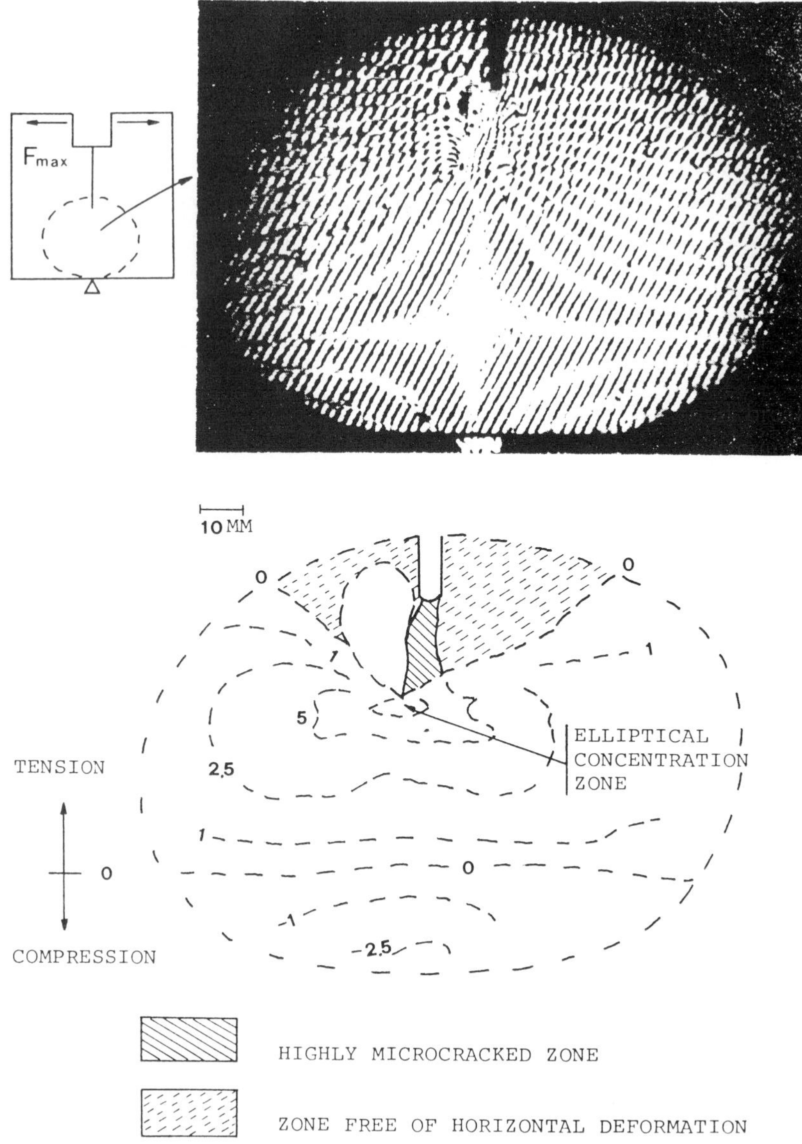

Fig. 11. Measurement of FPZ in mortar specimen with holographic interferometry (after [23]).

or size dependent [8, 15]. It has been proved [7, 15] that

$$K_r = \alpha \cdot K_0 \cdot \frac{G_f}{f_t} \cdot \frac{aC'(a)}{C(a)[C(a) - C(a_{\text{eff}})]},$$

$$= \alpha \cdot K_0 \cdot \frac{G_f}{f_t w_c} \left\{ X \cdot \frac{C'(a)CMOD_u}{C(a)[C(a) - C_u]} \right\}, \quad \alpha \leqslant 1. \tag{19}$$

This equation reveals the intrinsic connection among the fracture toughness, FPZ and the fracture energy G_f. The dependence of K_r on the compliance C explains why there is a size effect if FPZ is not small compared with the dimensions of the specimen. A size effect in the fracture toughness has also been observed in hardened cement paste [24] and discussed in [25].

The distribution of the total fracture energy over a fracture area, of course, is affected by FPZ. If the variation in the length of a fully-developed FPZ is mainly attributed to the stress gradient ahead of a traction-free crack, it is expected that the width of FPZ will increase or decrease with the length. Therefore, a size effect in fracture energy G_f will be inevitable due to the variation in FPZ. It is particularly true if some artificial efforts, such as deep grooves in thin specimens, have not been made. The results of Fig. 10 imply that further investigation on the width of FPZ and its influence upon G_f is imperative.

The low G_f value of 21.8 N/m obtained from mortar A could be another indication of the width influence of FPZ upon G_f. The width of FPZ for this material with the maximum sand of 4 mm could be as wide as 12 mm (for concrete) according to CBT. However, the 2 mm wide grooves with a depth of 2 mm on the thin specimens (of 15 mm in thickness) largely limit such a possibility. The thickness of the specimens also contributes to the low G_f value [26]. For the same specimen geometry and dimension, G_f of mortar B with the maximum sand of 1 mm is 33 N/m. In this case, the width of FPZ, possibly 3 mm on an average, is less influenced by the deep grooves.

6. Conclusion

A major feature of this work summarized from our previous investigations [5–8] is the simplicity and generality of both the experimental and the analytical methods. With the methods presented in this paper, the measurement of FPZ, the bridging stress and the strain-softening becomes relatively easy. Furthermore, a prediction of the extension of FPZ is feasible. A complete picture on the extension of FPZ during fracture is particularly helpful in the understanding of the K_r-curve behaviour of concrete-like materials. The good agreement between experimental results and theoretical predictions on the length of FPZ implies the potential to construct K_r curves for either the traction-free crack propagation or the total crack propagation. K_r curves based on these two crack propagations will certainly be more objective than a K_r curve for an effective crack.

The compliance curve C_p from the stepwise cutting procedure is of significance in understanding the role of FPZ in fracture toughness and fracture energy. Equations derived from C_p [7, 15] reveal the intrinsic relations among the extension of FPZ and the strain-softening. The multi-cutting technique has been successfully proved with a numerical simulation [22]. Results of the computer simulation show that the strain-softening curve derived from C_p is reliable, and the manner of compliance evaluation described in [5, 15] is adequate.

Experimental results on FPZ indicate that the length of FPZ is not constant during fracture and that further investigation on the width of FPZ is merited in order to fully understand the size dependence of G_f and K_{IC}.

References

1. A. Hillerborg, M. Modeer and P.E. Petersson, *Cement and Concrete Research* 6 (6) (1976) 773–782.
2. A. Hillerborg, in *Fracture Mechanics of Concrete*, F.H. Wittmann (ed.), Elsevier, Amsterdam (1983) 223–249.

 3. Z.P. Bazant and L. Cedolin, *Journal of the Engineering Mechanics Division*, ASCE 105 (1979) 297–315.
 4. Z.P. Bazant and B.H. Oh, *Materials and Structures*, RILEM 16 (1983) 155–177.
 5. X.Z. Hu and F.H. Wittmann, *Journal of Materials in Civil Engineering* (1990).
 6. X.Z. Hu and F.H. Wittmann, in *Fracture of Concrete and Rock*, S.P. Shah, S.E. Swartz and B. Barr (eds.), Elsevier, England (1989) 307–316.
 7. X.Z. Hu and F.H. Wittmann, 'An Analytical Method to Determine the Bridging Stress Transferred within the Fracture Process Zone: I, General Theory', submitted to *Cement and Concrete Research* (1990).
 8. X.Z. Hu and F.H. Wittmann, 'An Analytical Method to Determine the Bridging Stress Transferred within the Fracture Process Zone: II, Application to Mortar', submitted to *Cement and Concrete Research* (1990).
 9. P. Nallathambi and B.L. Karihaloo, in *Fracture Toughness and Fracture Energy of Concrete*, F.H. Wittmann (ed.), Elsevier, Amsterdam (1986) 271–280.
10. A.S. Kobayashi, N.M. Hawkins, D.B. Barker and B.M. Liaw, in *Applications of Fracture Mechanics to Cementitious Composites*, S.P. Shah (ed.), Martinus Nijhoff Publishers, Dordrecht (1985) 25–50.
11. R. Knehans and R. Steinbrech, *Journal of Materials Science, Letters* 1 (1982) 327–329.
12. M. Sakai, J. Yoshimura, Y. Goto and M. Inagaki, *Journal of American Ceramic Society* 71 (8) (1988) 609–616.
13. X.Z. Hu and F.H. Wittmann, 'Fracture Energy and Fracture Process Zone', to be published (1990).
14. E. Bruhwiler and F.H. Wittmann, *Engineering Fracture Mechanics*, to be published.
15. X.Z. Hu, *Fracture Process Zone and Strain-Softening in Cementitious Materials*, Postdoctoral Research Report, Institute for Building Materials, Swiss Federal Institute of Technology, Zurich (1989).
16. RILEM Draft Recommendation, *Materials and Structures* 106 (1985) 285–290.
17. X.Z. Hu, B. Cotterell and Y.W. Mai, *Proceedings of the Royal Society* (*London*), A 401 (1985) 251–265.
18. X.Z. Hu, *Statistical Fracture of Brittle Materials*, Ph.D thesis, Department of Mechanical Engineering, Sydney University, Sydney (1988).
19. M.E. Raiss, J.W. Dougill and J.B. Newman, in *Fracture of Concrete and Rock*, S.P. Shah, S.E. Swartz and B. Barr, (eds.), Elsevier, Amsterdam (1989) 243–253.
20. T. Hashida, in *International Workshop on Fracture Toughness and Fracture Energy – Test Methods for Concrete and Rock*, Sendai, Japan (1988) 37–47.
21. R.M.L. Foote, Y.W. Mai and B. Cotterell, *Journal of Mechanics and Physics of Solids* 34 (6) (1986) 593–607.
22. A.M. Alvaredo, X.Z. Hu and F.H. Wittmann, in *Fracture of Concrete and Rock*, S.P. Shah, S.E. Swartz and B. Barr (eds.), Elsevier, Amsterdam (1989) 51–60.
23. Ph. Regnault and E. Bruhwiler, *Engineering Fracture Mechanics*, to be published.
24. D.D. Higgins and J.E. Bailey, *Journal of Materials Science* 11 (1976) 1995–2003.
25. B. Cotterell and Y.W. Mai, *Journal of Material Science* 22 (1987) 2734–2738.
26. S.L. Xu and G.F. Zhao, *International Workshop on Fracture Toughness and Fracture Energy – Test Methods for Concrete and Rock*, Sendai, Japan (1988) 157–163.

International Journal of Fracture **51**: 19–29, 1991.
Z.P. Bažant (ed.), Current Trends in Concrete Fracture Research.
© 1991 *Kluwer Academic Publishers. Printed in the Netherlands.*

Observation of fracture process zone by laser speckle technique and governing mechanism in fracture of concrete

HIDEYUKI HORII[1] and TOSHIMICHI ICHINOMIYA[2]
[1]*Department of Civil Engineering, University of Tokyo, Bunkyo-ku, Tokyo, Japan;*
[2]*Kajima Institute of Construction Technology, Kajima Corporation, Tobitakyu, Chofu, Tokyo, Japan*

Received 11 June 1990; accepted 11 November 1990

Abstract. Fracture tests with a wedge-loading device are carried out on mortar and concrete specimens so as to have stable crack growth. Using laser speckle technique the length of macrocrack and the distribution of crack opening displacement are measured. Results are compared with those obtained by the boundary element method (BEM) analysis for a Dugdale-Barenblatt-type model of a fracture process zone. The governing mechanism in fracture of concrete and the mechanism which is represented by the model are discussed with special attention to the microcracking zone. It is deduced that a Dugdale-Barenblatt-type model does not represent the microcracking zone, thus implying that the microcracking zone and the bridging zone correspond to the pre-peak nonlinear part of the stress-strain curve in a uniaxial tension test and the post-peak tension-softening curve, respectively. It is concluded that the effect of microcracking on the maximum load is less significant than that of bridging. Possible models which include the effect of microcracking in addition to that of bridging are proposed.

1. Introduction

Recently remarkable progress has been accomplished in the study of fracture mechanics of concrete and rock. The goal is to provide a mathematical description of fracture phenomena in those materials. To this end the identification of fracture mechanisms and their modeling is necessary. Recent progress in this field is the result of the observation of the fracture process zone and the studies on its analytical modeling [1, 2]. Observations of local phenomena near the crack tip with advanced techniques such as acoustic emission (AE), laser holographic interferometry, and moiré interferometry, have provided a better insight into the pertinent fracture mechanisms [3–6].

The fracture phenomena in concrete and rock are characterized by the existence of a fracture process zone at the tip of the crack. Major mechanisms in this zone are microcracking and bridging (stress transmission across macrocrack surfaces) by aggregates or fibers (Fig 1). A visible (macroscopic) crack along which stresses are transmitted should be included in the fracture process zone. The state inside the fracture process zone corresponds to a point on the tension-softening curve obtained from a uniaxial tensile fracture test. The tension-softening curve is considered to represent material characteristics of the fracture, and the Dugdale-Barenblatt-type model with a tension-softening curve†, which was first introduced by Hillerborg [7], may be the appropriate model for the fracture process zone in concrete and rock; the size effect law of Bažant [8], the two-parameter model of Shah [9], and

---

†The model has been named differently, e.g. Fictitious Crack Model, Cohesive Crack Model. Since the modeled crack is not fictitious and the stress transmitted across the bridging zone is not cohesive, the name Dugdale-Barenblatt-type model (with a tension-softening curve) is used in the present paper.

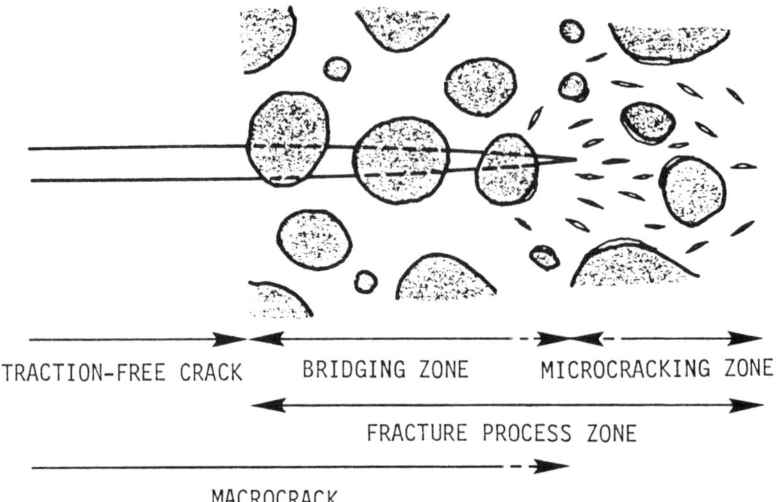

Fig. 1. Schematic illustration of fracture process zone which consists of microcracking zone and bridging zone.

the effective crack model of Karihaloo [10] produce similar results under usual conditions [10, 11].

Although a number of facts has been clarified, there are still more to be investigated. What is the role of microcracking and bridging on the toughness in different materials? Does the Dugdale-Barenblatt-type model represent the microcracking zone as well as the bridging zone? Is the macroscopic crack formed before, at, or after the peak stress? To study these points, direct observation and theoretical study of the fracture process zone must be combined.

In the present work, fracture tests of mortar and concrete specimen are carried out. A wedge-loading apparatus is used to obtain stable crack growth. The displacement on the specimen surface is measured using the laser speckle technique within an accuracy of a few micron (μm $= 10^{-3}$ mm). The length of the macrocrack and the distribution of crack opening displacement are measured at each loading stage. They are then compared with the analytical results of the Dugdale-Barenblatt-type model obtained by the BEM. Through the comparison, the validity of the model is discussed with particular attention to the microcracking zone.

2. Observation by laser speckle technique

2.1. Laser speckle technique

A laser speckle technique is a method to measure the surface displacement of a body using laser interferometry [6]. A laser beam is expanded by a lens and a pin hole, and the uniform light is applied on the specimen. The reflected light produces a random-shaped speckle pattern of unusually high contrast on the image formed by a lens. Each speckle is random-shaped but is almost of the same size. When the specimen moves, the speckle moves with the specimen.

We expose the speckle pattern twice on the same film before and after the deformation. By the double exposure, two sets of speckles, which are slightly apart corresponding to the magnitude of displacement at each point, are recorded on the film. After developing the negative film, we apply a laser beam at a point on the film. Then fringes appear on the screen. The direction of the fringe is perpendicular to the direction of the displacement, and the distance between the fringes is inversely proportional to the magnitude of displacement at the point where the laser beam is applied. Measuring the spacing and the direction of the fringe, the magnitude and direction of the displacement at the point are obtained. In this study the fringe pattern is recorded using a video camera and transformed into a digitized image which is taken into a personal computer. Using an image analysis technique and the fast Fourier transform, the spacing and the direction of the fringe are calculated (Fig. 2).

To check the accuracy, a rigid body displacement of a body was measured by laser speckle technique, and results are compared with those measured using a displacement transducer (strain gauge type). The difference is 1–2 μm for the displacement in the range of 250–300 μm. When the displacement is smaller than 50 μm, the fringe does not appear. When the displacement is larger than 300 μm, the spacing of the fringe is too close. The most adequate range of displacement is 250–300 μm for the present condition.

2.2. Fracture test

The configuration of the wedge-loaded specimen used in this study is shown in Fig. 3. Notch lengths are 5, 6 and 7 cm. Both mortar and concrete specimens were tested. The mix proportions of the castings by weight for mortar and concrete are shown below:

	Cement	Sand	Gravel	Water
Mortar	1.00	2.00		0.50
Concrete	1.00	2.36	1.80	0.50

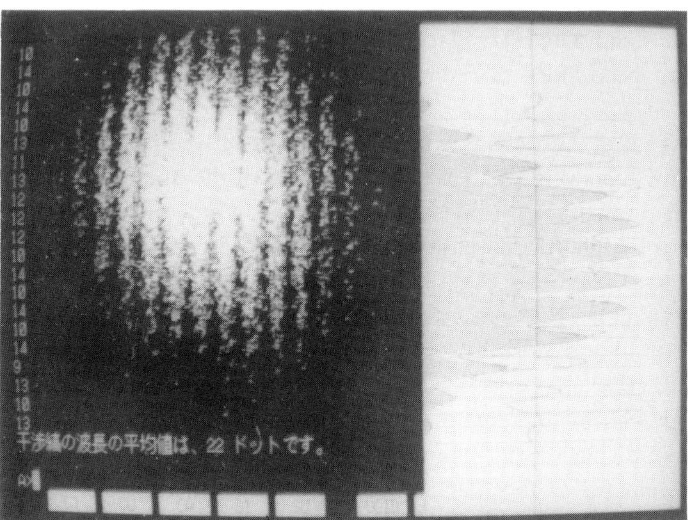

Fig. 2. Digitized image of the fringe pattern and results of image analysis on CRT.

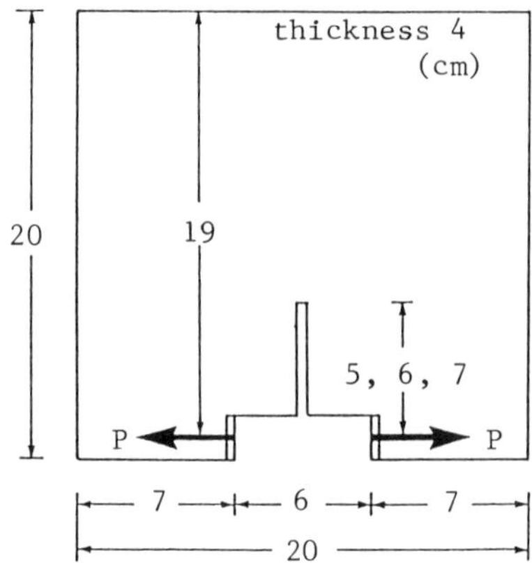

Fig. 3. Wedge-loaded specimen of mortar and concrete; steel plates are enbedded on which load is applied with a wedge-loading device.

Normal Portland cement was used, and the sand and gravel (crushed) were obtained locally with a maximum aggregate size of 1 cm. The specimens were cured in water for 28 days before the test; there were 12 mortar specimens and 16 concrete specimens. Each time 4 specimens were prepared with the same mix. The cylinder compressive strength is 40.5 MPa as the average (38.6–42.4 MPa; variation in average of 3 cylinder tests) for mortar and 43.8 MPa (41.3–46.4 MPa) for concrete. The cylinder splitting strength is 3.60 MPa (3.31–3.88 MPa) for mortar and 3.61 MPa (3.33–3.77 MPa) for concrete.

Before the test, a 6 mm spacing grid was drawn on the specimen. The grid is used to identify the point at which the displacement is measured by laser speckle technique. A small loading frame was prepared so that the test was performed in a darkroom. The loading frame was set on a steel plate mounted on cylindrical bars to give rigid body displacement. The displacement of the wedge is controlled and the axial force applied to the wedge was measured. The wedging load which was applied on the specimen through the wedge is converted from the axial force. Displacement transducers (strain gauge type) are fixed at the tip of the notch on both sides of the specimen to measure the opening displacement at the notch tip.

The loading procedure is to provide a small amount of wedge displacement and maintain it for a minute to allow relaxation. The average time rate of the opening displacement at the notch tip is controlled at about 0.7 μm/min as an average for mortar and 0.9 μm/min for concrete. When the speed is too high, the maximum load becomes larger, and after the peak load unstable crack growth occurs. After the peak load the increment of wedge displacement applied is decreased to a very small value close to zero.

The first of the double exposure is made before we start loading; the second exposure on the first film is taken when the notch tip opening displacement reaches 8–10 μm. Then the process is repeated with an increment of 5–7 μm in the notch tip opening displacement. The appropriate range of displacement magnitude for the laser speckle technique is 200–300 μm. To keep all the displacements of the specimen in this range, we apply rigid body displacement to the loading frame on which the specimen is mounted when the second one of the double exposure is made; the loading frame is set on a steel plate on cylindrical bars. To measure the macrocrack length and the distribution of crack opening displacement, only the relative displacement is required.

The specimen is fractured into two parts after the test in order to identify the crack path referring to the grid drawn on the specimen. After developing the negative films, we reflect the laser beam on them to produce the fringe pattern and obtain the displacement at each point following the procedure explained previously. The displacements at just above and below the crack path are measured along the path; the distance between the two points is about 5 mm. Crack opening displacement is obtained as their difference. The location of the macrocrack tip is interpreted to be the point of vanishing crack opening displacement. In this manner, the length of the macrocrack and the distribution of crack opening displacement increments are obtained at each loading stage.

2.3. Experimental results

The relation between the load and the notch tip opening displacement is shown in Fig. 4 for mortar and concrete specimens with notch lengths of 5, 6 and 7 cm. Twelve mortar specimens and 16 concrete specimens show that results are reproducible up to just after peak loads. The notch tip opening displacements measured at both sides match each other. As the notch tip opening displacement increases, the macrocrack emanates from the notch tip and grows in a stable manner. The macrocrack can be observed with the naked eye when the notch opening displacement is about 50 μm. The double exposures are made at each loading stage shown in Fig. 4 by circular points. The crack opening displacement obtained from each negative film is the increment during the double exposure. The total value of crack opening displacement at each loading stage is obtained by summing all the preceding increments.

The distribution of total crack opening displacement and the length of the macrocrack are shown in Fig. 5 for a mortar specimen with notch length 6 cm and for a concrete specimen with notch length 7 cm.

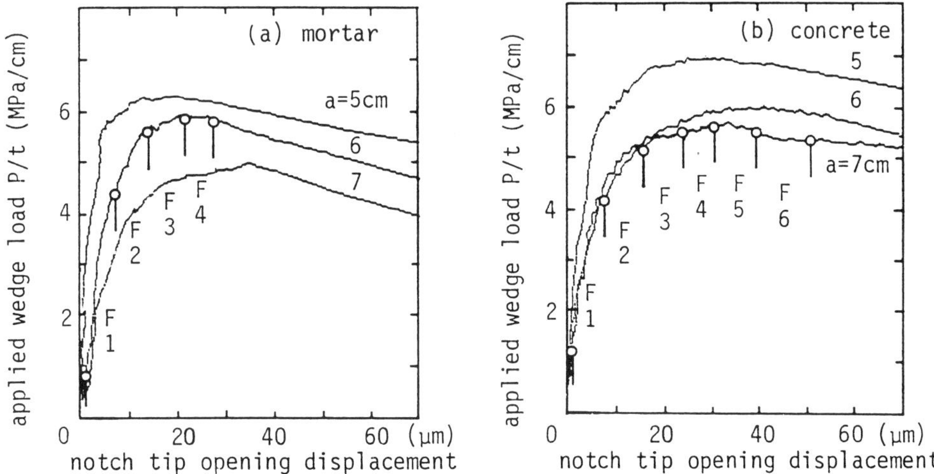

Fig. 4. Relation between applied wedge load (per unit thickness) and notch tip opening displacement obtained from displacement transducers; F1, F2,... indicate intervals of double exposures on film 1, film 2,....

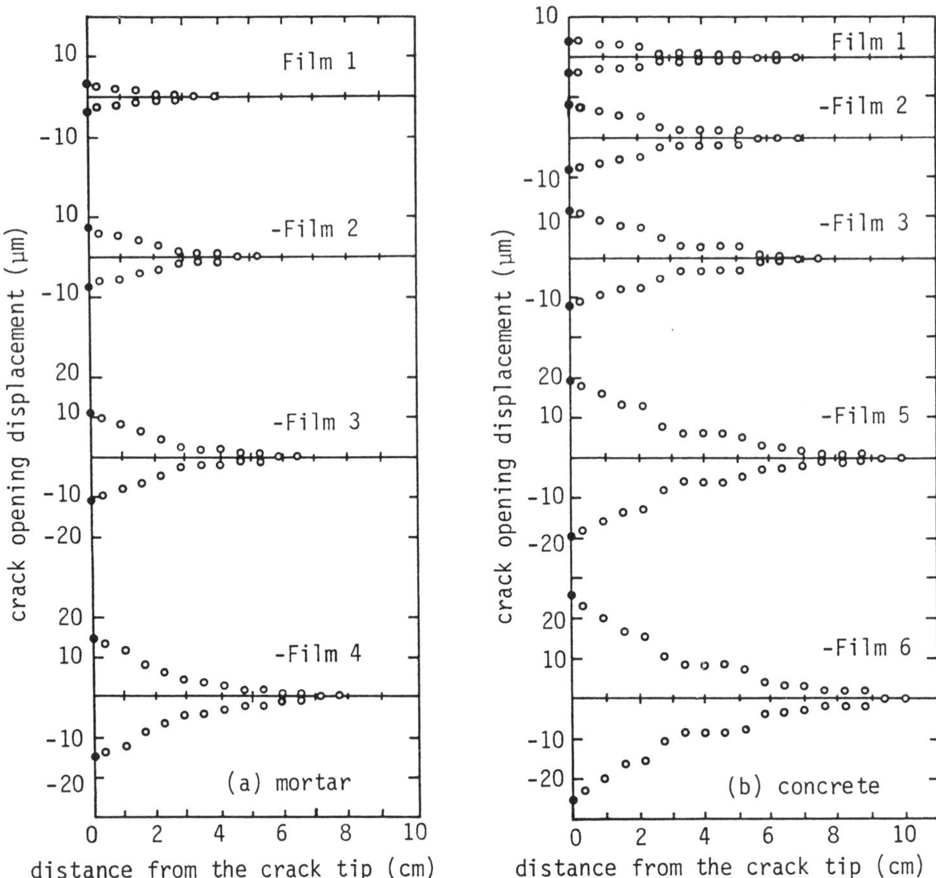

Fig. 5. Variation of crack opening displacement at each loading stage.

3. BEM analysis of Dugdale-Barenblatt-type model

The Dugdale-Barenblatt-type model with a tension-softening curve is considered to be an appropriate model of the fracture process zone and is considered to represent mainly the bridging zone where the stresses are transmitted by aggregates or reinforcing fibers. It is yet to be discussed whether or not the Dugdale-Barenblatt-type model represents the microcracking zone as well.

In the present study a simple linear tension-softening relation is employed. That is, the relation between the transmitted stress σ and the crack opening displacement w is given by

$$\sigma = f_t(1 - w/w_c), \tag{1}$$

where f_t and w_c denote the tensile strength and critical opening displacement, respectively. Note that we consider a linear line which is tangent to the real tension-softening curve at $\sigma = f_c$. For a specimen as small as the one used in the present study, the maximum opening displacement is small and this approximation is valid up to approximately the peak load.

At the end of the extended crack tip it is assumed that:

$$K_{\mathrm{I}} = 0, \tag{2}$$

where K_{I} is the stress intensity factor of mode I. This condition is based on the idea that the stress is relaxed by the formation of the fracture process zone and the stress singularity does not exist. This condition is closely related to modeling of the microcrack formation as discussed later.

We have employed the BEM (or boundary integral equation method, BIEM) to satisfy the conditions (1) and (2) as well as the conditions along the outer boundary and traction free crack. The boundary value problem is transformed into an integral equation for unknown quantities on the boundary based on Green's formula and properties of the fundamental solution. By discretizing the boundary quantities, the boundary integral equations are reduced to a system of linear algebraic equations which is then solved numerically.

Since the unknown quantities are defined only on the boundary, BIEM is suitable for the analysis of crack growth problems. For crack problems, however, a special formulation is necessary for the crack boundaries since four boundary quantities, the displacement and the traction on the upper and lower surfaces are defined along the crack boundary, and the values of two linear combinations of them are prescribed as the boundary conditions. Horii [12] showed a general formulation for crack problems with arbitrary boundary conditions which can be easily applied to the Dugdale-Barenblatt-type model. The introduction of conditions (1) and (2) in the BEM program is straightforward.

In the present calculations the input data of the material are f_t, w_c, and Young's modulus E. A value of $f_t = 3.92$ MPa was used. (For comparison results of the simple model [13] is also shown with $f_t = 3.43, 3.92, 4.41$ MPa.) The value of w_c is set to be 25 μm referring to the reported tension-softening curve obtained from direct tension tests [14]. Note that the linear tension-softening relation considered here is the one tangent to the real tension-softening curve at $\sigma = f_t$. $E = 3.26 \times 10^4$ MPa is used for the Young's modulus.

4. Comparison and discussion

The analytical and experimental results are shown in Fig. 6. Results of the simplified model [13] are also shown for comparison. The simplified model is a Dugdale-Barenblatt-type model with a linear tension softening relation at the tip of a semi-infinite crack in an infinite plane. The effect of specimen geometry and loading condition is taken into account by a single parameter λ representing the second order term in the near tip stress distribution.

A reasonable agreement between experimental results and the Dugdale-Barenblatt-type model is obtained for the maximum load. Notice that the observed length of the macrocrack is larger than the predicted values by BEM analysis. The difference is larger for the concrete specimen than the mortar specimen. This point is discussed later. The results of the simplified model [13] show longer plateaux, due to the assumption of a semi-infinite crack in an infinite plane. In BEM analysis, for a macrocrack longer than the one shown in Fig. 6, the maximum opening displacement reaches its critical value thus implying the full growth of bridging zone. This may be different from reality because of the assumption of a linear tension-softening

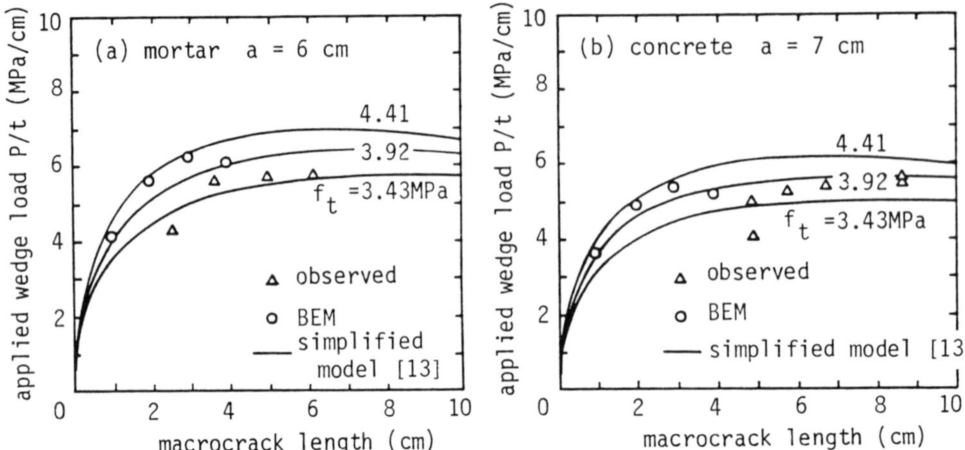

Fig. 6. Relation between applied wedge load (per unit thickness) and length of macrocrack.

relation. The linear tension-softening relation used in this study is the one which is tangent to the real curve at the peak point. The long tail of the real tension-softening curve may result in the slow decay of the load after the peak, but it may not much affect the value of peak load.

The distribution of crack opening displacement for mortar is compared with that obtained by BEM analysis in Fig. 7; they agree well except near the crack tip. The observed macrocrack (or the displacement discontinuity) is longer than the prediction by the model. This requires careful consideration and discussion since it is closely related to the fundamental question as to how well the model represents the actual phenomena in the fracture process zone.

Several interpretations of this difference are possible.

1. It is due to the error in measurement by the laser speckle method. There is no displacement discontinuity in that part. The magnitude of opening displacement (or the

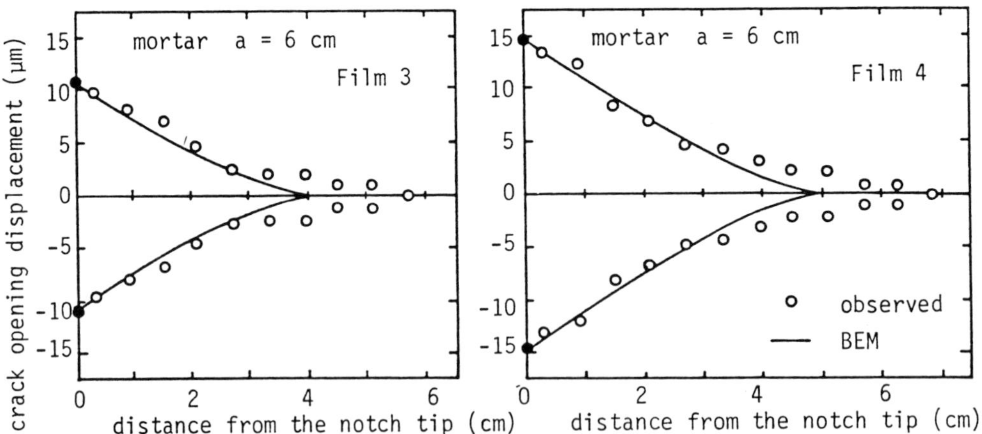

Fig. 7. Variation of observed and predicted crack opening displacement; closed points present notch tip opening displacements obtained from displacement transducers.

difference in displacements between the upper and lower sides of the crack path) is a few µm which is of the same order as the accuracy of the measurement by the laser speckle technique.

2. The measured difference in displacement is not due to the crack opening but due to the deformation induced by microcracking. The distance between the two points where the displacement is measured is about 5 mm. Because of the microcracking, the deformation is larger than the elastic deformation, and it was detected in the measurement.

3. The difference is due to the actual displacement discontinuity. The macrocrack is longer than that predicted by the model.

Other possibilities, such as the difference between surface and inside, are also considered. To clarify the source of difference further observation with accurate measurement is necessary. It is not possible to specify the situation at this moment. However, for each case we may be able to deduce the following conclusions. For the first case, there exist no problems except for the measurement by laser speckle technique. The Dugdale-Barenblatt-type model with tension-softening curve describes the fracture phenomena in concrete. For the second case, there exists a measurable microcracking zone and it is not represented by the Dugdale-Barenblatt-type model. And for the third case, the Dugdale-Barenblatt-type model does not predict the macrocrack length correctly. The model needs to be modified to predict the accurate macro-crack size. This implies that in a uniaxial tension test, the macrocrack is formed before the peak-point because the tension-softening curve, on which the model is based, represents material behavior after the peak.

In all cases, it is deduced that the Dugdale-Barenblatt-type model does not represent the microcracking zone. If the model includes the microcracking zone, the predicted length of the fracture process zone must be longer than the observed macrocrack length. If the measurement includes deformation due to microcracking as in case 2, the predicted length and observed one may be equal; it, of course, depends on the extent to which the microcracking is involved in the model or in the measurement.

Since the maximum load and the distribution of the crack opening displacement predicted by the model agree with experimental observations, it may be concluded that the effect of microcracking is less significant in the fracture of concrete. To check the effect of the microcracking, further experimental observations and examinations of an analytical model which includes the effect of microcracking are required.

Case 3 may be excluded, since there is no experimental support which shows that the macrocrack (fracture plane) is formed before the peak stress in the uniaxial tension test. If the first case holds, it is not necessary to improve the fracture process zone model. If the second case holds, the fracture phenomena may be described as follows: The microcracking zone corresponds to the nonlinear part before the peak in the uniaxial tension test. At the peak stress the macrocrack is formed, and the bridging, which is the stress transmission across the macrocrack surface, corresponds to the tension-softening curve after the peak. We may accept the present Dugdale-Barenblatt-type model with the excuse that the effect of microcracking zone is less significant.

If we want to modify the present model and include the effect of microcracking, the following three choices are considered:

1. Instead of the condition (2), i.e. a zero stress intensity factor, assume

$$K_{\mathrm{I}} = K_{\mathrm{Ceff}}, \tag{3}$$

where the effective fracture toughness K_{Ceff} represents the effect of microcracking. Whether this treatment for microcracking is applicable should be examined.

2. Instead of the tension-softening curve which represents the post-peak behavior, a relation including pre-peak nonlinear behavior is used. Since pre-peak behavior is not a fully localized behavior, some means for the translation from the stress-strain curve to the transmitted stress vs. opening displacement curve is required.

3. The direct model of the microcracking zone is introduced in the model not by the extended part of the macrocrack. Then the model includes both the bridging zone and the microcracking zone. This seems to be most realistic and to provide the clarification of rules for both bridging and microcracking. However, the model becomes complicated and requires extensive numerical computation.

In any model, the comparison with accurate observation is necessary to verify its validity.

5. Summary

In the present study the length of macrocrack and the distribution of crack opening displacement are measured using the laser speckle technique. Results are compared with those of the Dugdale-Barenblatt-type model with a linear tension-softening relation. A reasonable agreement is obtained on the maximum load.

The observed and predicted distributions of crack opening displacement agree except for a part near the crack tip. The observed displacement discontinuity is longer than the prediction. From this difference it is deduced, leaving aside the discussion on the accuracy of the measurement, that the Dugdale-Barenblatt-type model does not represent the microcracking zone and the effect of microcracking is less significant on the maximum load.

The present study illustrates the necessity for accurate experimental observations and an analytical model of the fracture process zone which models microcracking as well as bridging.

Acknowledgments

This paper was originally prepared for the *Proceedings of the International Conference on Recent Developments in the Fracture on Concrete and Rock*, Cardiff, UK, Sept. 1989, and was completed after its dead line. The authors would like to thank its editor, Professor B. Barr, who encouraged us to publish this paper elsewhere. This study was supported in part by the Grant-in-Aid for Scientific Research, from the Japanese Ministry on Education, Science, and Culture. The authors wish to thank M. Ueno, S. Fujii, T. Tokunaga, and T. Niijima (University of Tokyo) for their help in the experiments and computations.

References

1. *Fracture of Concrete and Rock, Proceedings of SEM–RILEM International Conference*, Jun. 17–19, 1987, Houston, S.P. Shah and S.E. Swartz (eds.), Springer-Verlag (1989).
2. *Cracking and Damage: Strain Localization and Size Effect, Proceedings of France–US Workshop on Strain Localization and Size Effect due to Cracking and Damage*, Sept. 1988, Cachan, France, J. Mazars and Z.P. Bažant (eds.), (1989).

3. A.K. Maji and S.P. Shah, in *Nondestructive Testing*, SP-112, H.S. Lew (ed.), ACI (1988) 83–109.
4. R.A. Miller, S.P. Shah and H.I. Bjelkhagen, *Experimental Mechanics* (1988) 388–394.
5. J.J. Du, A.S. Kobayashi and N.M. Hawkins, in [1] (1989) 199–204.
6. P. Jacquot and P.K. Rastogi, in *Fracture Mechanics of Concrete*, F.H. Wittmann (ed.), Elsevier (1983) 113–155, Chapter 3.4.
7. A. Hillerborg, M. Modéer and P.E. Petersson, *Cement and Concrete Res* 6 (1976) 773–782.
8. Z.P. Bažant, *Journal of Engineering Mechanics*, ASCE 110 (1984) 518–538.
9. Y.S. Jenq and S.P. Shah, *Journal of Engineering Mechanics*, ASCE 111 (1985) 1227–1241.
10. B.L. Karihaloo and P. Nallathambi, Final Report of Sub-committee A Notched Beam Test; Mode I Fracture Toughness of RILEM TC89-FMT (1988).
11. J. Planas and M. Elices, in [2] (1989) 462–476.
12. H. Horii, in *Boundary Element Methods in Applied Mechanics, Proceedings of the First Joint Japan/US Symposium on Boundary Element Methods*, Tokyo, Oct. 3–6, 1988, M. Tanaka and T.A. Cruse (eds.), Pergamon Press (1988) 129–138.
13. H. Horii, A. Hasegawa and F. Nishino, in [1] (1989) 205–219.
14. H.A.W. Cornelissen, D.A. Hordijk and H.W. Reinhardt, in *Proceedings of Conference on Fracture Mechanics of Concrete*, 1, E.P.F.L., Switzerland (1985) 419–429.

International Journal of Fracture **51**: 31–42, 1991.
Z.P. Bažant (ed.), Current Trends in Concrete Fracture Research.

Tensile fracture of concrete at high loading rates taking account of inertia and crack velocity effects

H.W. REINHARDT[1] and J. WEERHEIJM[2]
[1]Stuttgart University, Pfaffenwaldring 4, D-7000 Stuttgart 80, Germany;
[2]TNO Prins Maurits Laboratory, P.O. Box 45, NL-2280 AA Rijswijk, Netherlands

Received 1 June 1990; accepted 15 November 1990

Abstract. Experiments on concrete have shown a considerable increase of apparent tensile strength with increasing loading rate. This phenomenon can be attributed to various mechanisms such as kinetics of energy barriers and inertia effects in the vicinity of a running crack. After a general introduction into the behaviour of concrete under a uniaxial tensile stress, a model is presented which accounts for flaws in concrete and the inertia effects around a crack at high loading rates. It turns out that the quite crude model predicts rather accurately the relation between apparent strength and loading rate, the influence of concrete quality on this relation, and the relation between crack propagation velocity and loading rate.

1. Introduction

Civil engineering structures are usually exposed to static loading or slowly varying loading. However, there are loading cases with high rates of loading such as impact of vehicles or planes on structures, earthquake loading, explosions and hazards. Although these are the exceptional loading cases, rather than the rule, they may be vital for the structure and for the inhabitants or users. Therefore, strength and fracture energy at high loading rate are important parameters in assessing the safety of structures.

Tensile behaviour of concrete controls cracking, minimum reinforcement ratio, shear failure of beams and slabs, bond deterioration and bond slip of steel bars in concrete, and local splitting failure. As has been shown by John and Shah [1], loading rate may be responsible for the transition of ductile bending type failure to brittle shear type failure of beams.

Although tensile strength is only one parameter which governs failure of concrete, it is a dominant one. It controls the onset of a fictitious crack according to Hillerborg [2]. According to this model, a tensile member is stressed up to the maximum stress, then deformation continues while the stress decreases. The softening causes elastic stress release of the uncracked part of the tensile member and irreversible deformation of a crack process zone. At complete stress decay, the crack borders are separated and a discrete crack has formed.

The area under the complete stress-displacement curve represents the fracture energy G_F. The ascending part of the stress-deformation line is linear to about 0.6 times the tensile strength whereafter an overproportional increase of deformation appears. The descending part is nonlinear. The shape of the line depends on the type of material and on the loading conditions [3]. G_F can be described by a function of crack opening w with the scale factor f_t which is the tensile strength

$$G_F = f_t g(w), \tag{1}$$

where g may be a bilinear or multilinear, or a continuous function of crack opening w.

In experiments on plain concrete and on steel fibre reinforced concrete, it has been shown that $g(w)$ is little affected by loading rate [4]. If $g(w)$ is taken as approximately the same for all loading rates, the rate dependence of G_F is the same as for the tensile strength. It is then sufficient to determine the tensile strength of concrete at various loading rates and to measure the complete stress-deformation curve at static loading.

2. Some aspects of fracture at high loading rates

The aforementioned relation between G_F and f_t is true for one single crack. This implies that, at high loading rates, also a single crack may occur. If multiple cracks form in the measuring length, the energy must be larger than in the case of a single crack.

There are at least three aspects which require attention: first, the bond breaking process, second, the inertia effect of the material adjacent to the crack, and third, the crack propagation velocity.

The breaking of bonds has been described by the kinetics of energy barriers (see for instance [5]). It is assumed that there is a bond breaking rate and a bond healing rate. If there is a mechanical stress applied, the breaking rate exceeds the healing rate. This causes time dependent deformation and/or cracking. Mihashi and Wittmann [6] have applied this model to concrete and found good correlation between model predictions and experiments. The rate dependence is expressed there by a power function

$$f_t/f_{t,0} = (\dot{\sigma}/\dot{\sigma}_0)^n, \tag{2}$$

with f_t and $f_{t,0}$ tensile strength at high loading rates and static loading, respectively, and $\dot{\sigma}$ the appropriate stress rates. n is a material constant with values between 0.03 and 0.06. According to this theory, the mechanism which leads to fracture is the same for all loading rates. Rigorous studies have shown that the power function expression is only valid for a moderate crack velocity [5], while higher velocities should be described by a more rigorous application of kinetics theory.

If two pieces of material are separated by a crack, they move away from each other with a certain velocity. The velocity depends on the crack propagation velocity, the compliance and mass density of the material, on the stress and, at varying stresses, on the loading rate. Kipp et al. [7] have shown that a rate dependence of tensile strength exists like

$$f_t = \beta \dot{\varepsilon}^{1/3}, \tag{3}$$

with β a material constant which depends on Young's modulus, critical stress intensity factor, and shear wave velocity. In fact, (3) is based on the effect of inertia of a single flaw under high strain rates. The applicability of the formula is confined to strain rates greater than $10^2\,\mathrm{s}^{-1}$ for concrete assuming a flaw size of 5 mm.

Crack propagation velocity is theoretically limited to about the Rayleigh wave velocity since this is the velocity of a wave which transports energy along a surface, for instance a crack surface. However, even under high strain rates, there have been no values measured which reach this terminal velocity ([8] and [9]). While [8] reports values of about 200 m/s, [9] reports velocities of about 1000 m/s. It should be noted, however, that the value mentioned last is determined in compressive loading tests with a less sensitive method than has been used in [8]. It seems at the moment that the crack propagation velocity of concrete is considerably lower than the Rayleigh wave speed.

If this is true, there should be an influence on the tensile strength at very high loading rates when the loading increases continuously but the crack cannot propagate fast enough. Theoretically, the apparent strength could rise without limit. Curbach [10] performed tests and carried out finite element analyses assuming a crack propagation velocity of 500 m/s. He found a steep increase of tensile strength at strain rates greater than 1 s^{-1}. At $\dot{\varepsilon} = 50 \text{ s}^{-1}$ the theoretical ratio between tensile strength and static tensile strength amounted to 5.

In the following section, an attempt will be made to close the gap between the three aspects which seem to be independent from each other. This model will incorporate kinetic theory and inertia effects and will predict tensile strength as a function of loading rate, and crack velocity as a function of loading time and loading rate.

3. Concrete model

Failure of concrete is determined by the fracture process. In the model the fracture process, which occurs at micro level, is characterized by the extension of schematic cracks in a fictitious fracture plane. The influence of the loading rate on the material response and the ultimate stress is reflected in crack extension and energy absorption during the pre-peak fracture process.

3.1. Schematization

The real fracture plane is represented as a plane containing uniformly distributed penny-shaped cracks. These cracks represent the damage before loading. The ratio of the initial crack diameter and the intermediate distance (a_0/b) characterizes the ratio of damaged and undamaged material (for detailed description see [11]). Figure 1 shows the planes of failure in concrete and in the corresponding model.

The geometry of the fictitious failure plane is derived from the static material properties, water cement ratio, the aggregate content along with the general feature that bond cracking starts at about 60 percent of the strength and matrix cracking at about 75 percent. During the load increase the cracks grow and dissipate the fracture energies of bond, matrix and aggregate. When the crack size equals the intermediate distance b, the strength is assumed to be reached.

The only two parameters that cannot be derived from (static) material properties and therefore must be estimated are the specific surface energies for bond and aggregates.

In the fictitious failure plane, the material between the flaws is assumed to be linear elastic. Consequently, crack extension in this plane does not describe the processes of crack arrest, crack branching or bridging of cracks. But it can reflect the influence of loading rate and geometry on average crack growth, energy absorption and thus the influence on strength.

3.2. Crack extension

During crack extension, an exchange of energy occurs between deformation energy, external work, dissipated fracture energy and kinetic energy in the material around the crack tip. The

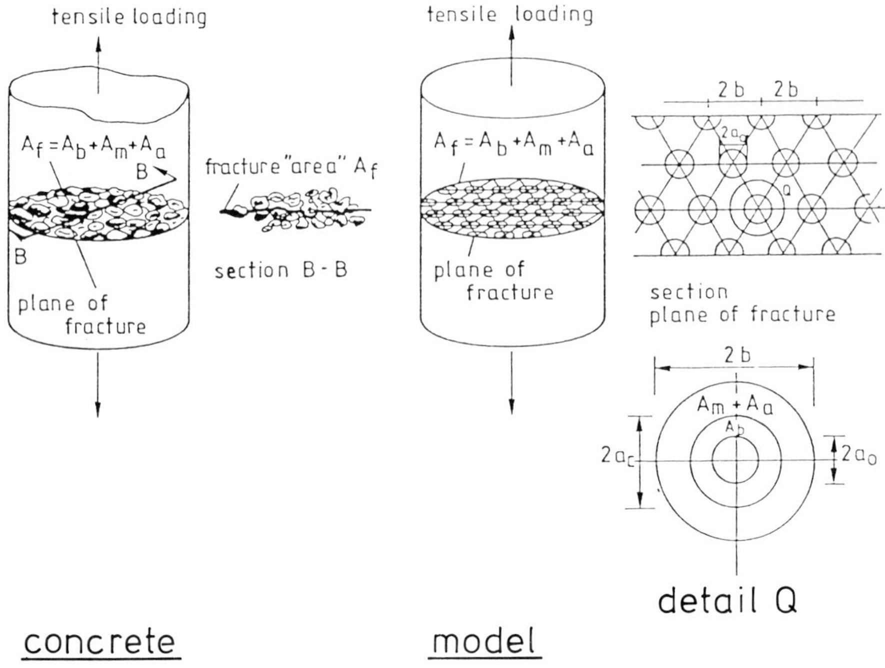

Fig. 1. Planes of fracture in concrete and model. A_b = bond area, A_m = matrix area, A_s = aggregate area, A_f = fracture area.

various energy terms can be determined if the stress, displacement and velocity fields are known. To determine these fields for dynamic loading conditions or at crack propagation, the equations of motion should be applied.

The effect of the inertia terms on the stress distribution is amongst others studied by Freund [12] and Broberg [13]. For a semi-infinite crack loaded by a longitudinal tension wave, Freund derived an expression for the dynamic stress intensity factor. The static factor should be multiplied by a function $k(\dot{a})$, which depends only on the crack velocity. It equals unity for $\dot{a} = 0$, and is reduced to zero at the limiting (Rayleigh) wave velocity (C_r).

For the geometry in the fictitious plane no analytical solution is available, but [14] and [15] showed that also for other crack geometries the velocity of crack extension is limited by the Rayleigh or shear wave velocity. Considering the force equilibrium around a crack tip this result can be argued. The reasoning is as follows. For any crack geometry the stress intensity factor will decrease when the crack tip velocity increases because the inertia components in the equilibrium become more important. A lower stress intensity means a lower energy flux into the fracture zone which causes a decrease of crack extension velocity. For any crack geometry, there must be a terminal velocity of crack extension with an equilibrium between the energy flux into the fracture zone and the energy release rate. When the terminal velocity equals C_r, the stress intensity factor must be zero because at this velocity the material around the newly formed crack surfaces cannot respond to the actual crack size and the equilibrium is realized by inertia forces. Based on these results, it can be assumed that also for the penny-shaped cracks a terminal velocity exists and that the stress intensity factor will decrease with increasing crack tip velocity.

Another aspect is that for the fictitious fracture plane the influence of the bounded geometry on the stress fields under dynamic loading is not as yet known. However a few comments are in order. First of all it must be true that also for a bounded geometry the crack tip velocity will never exceed C_r, and the dynamic stress intensity factor K_{ID} will be zero at this velocity because of the inertia effects. Second, from analytical solutions for step loadings on stable cracks, it follows that, after a certain time, the stress distribution can be described by the corresponding static stress field. The magnitude of this delay time depends on the crack size, the intermediate crack distance and the wave velocity C_r. For linear increasing loads, this result leads to a delay time in the response to the actual load level. When the time delay is supposed to be zero, the error introduced will decrease with decreasing intermediate distance and increasing fracture time, e.g. Nozaki et al. [16].

The aim of the model is not to give an exact description of the prepeak fracture process, but to characterize as simply as possible the crack extension and energy dissipation before the ultimate strength is reached. Based on this goal and the aforementioned features of crack extension, it was decided to neglect, at first, the time delay and to use the solutions from LEFM for the stress and displacement fields under static loading which are adjusted by a function $k(\dot{a})$, like Freund and Broberg [12, 13] derived, to account for inertia.

In this way, relatively simple expressions for the various energy terms are obtained and the dependency of the rate sensitivity on the various parameters can be examined. Using these adjusted solutions, the distribution of the available energy during the fracture process can be determined.

Corresponding to the approach of Mott [17] and Berry [18] and others, one considers the material around the crack tip in which the energy exchange occurs. For the penny-shaped cracks the material is bounded by the surface of a torus with radius R. The value of R is not known but should be proportional to a characteristic length parameter (see Fig. 2).

Following the approach of Mott and Berry, the integrated form of the energy balance is used:

$$-W + V + T + D = E_0, \tag{4}$$

in which E_0 equals the amount of energy inside the surface at the considered initial time t_0. W = external work, V = deformation energy, T = kinetic energy, and D = fracture energy.

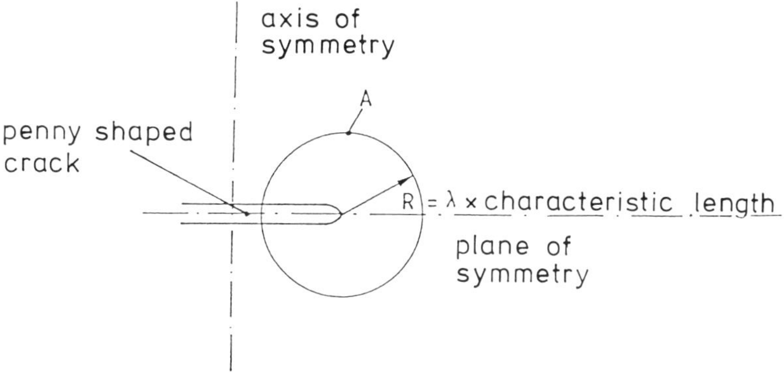

Fig. 2. The area around the crack tip. A = area boundary, λ = constant, R = radius of considered area.

36   *H.W. Reinhardt and J. Weerheijm*

Using the static solution from LEFM, expressions for the energy terms can be derived. (See Appendix). Substitution leads to the result that the balance is a function of load, crack size, crack velocity, known constants and one unknown proportionality constant λ.

Presentation of the mathematical formalism and discussion of the various energy terms exceeds the scope of this paper. These are given in [19]. To indicate the nature of the final result for the energy balance, it can be written, like the result of Mott and Berry, as a quadratic equation of $\dot{a}$:

$$AA(t)[\dot{a}(t)]^2 + BB(t) + \dot{a}(t) + CC(t) = 0. \tag{5}$$

In this equation the first two terms are mainly determined by the kinetic energy.

To determine the unknown constant λ, the analytical expressions for the work $W(t)$ and the deformation energy under static loading $V(t)$ of Sneddon [20] are used. λ equals 0.56 for a Poisson's ratio of 0.25. Assuming that the size of the fracture zone is not affected by the loading rate, the crack extension can be determined by solving (5).

Some results are presented in Figs. 3 and 4 [19]. Figure 3 shows the crack growth of one single crack, in case the stress fields are adjusted by the function $k(a)$ or not. The crack growth in the fictitious plane for various loading rates is given in Fig. 4.

From these results it emerges that for low loading rates the limiting velocity for a single crack is the Rayleigh wave, as it should be, but that the terminal crack velocity decreases with increasing loading rate. When the finite geometry of the fictitious fracture plane is considered, the same feature is observed for the rate dependency (Fig. 4). Calculations showed that the velocity decreases with decreasing intermediate crack distance. This aspect emerges also from a comparison of Figs. 3a and 4a.

3.3. Discussion of results

From the results presented in the former section, it is concluded that at least for low loading rates the crack extension model gives reliable results as the terminal crack tip velocity equals C_r. For higher loading rates the result is not directly obvious. An analysis of the effect of the loading

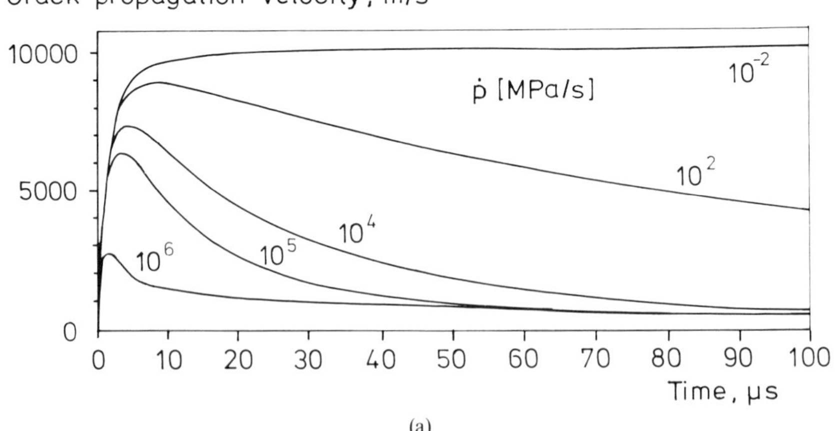

Fig. 3. Velocity of crack extension for various loading rates $\dot{p}$ for one crack. $\dot{p} = 10^{-2}$, 10^2, 10^4, 10^5, $5 \cdot 10^5$, 10^6 MPa/s. (a) No adjustments for inertia effects. (b) Adjustments for inertia effects.

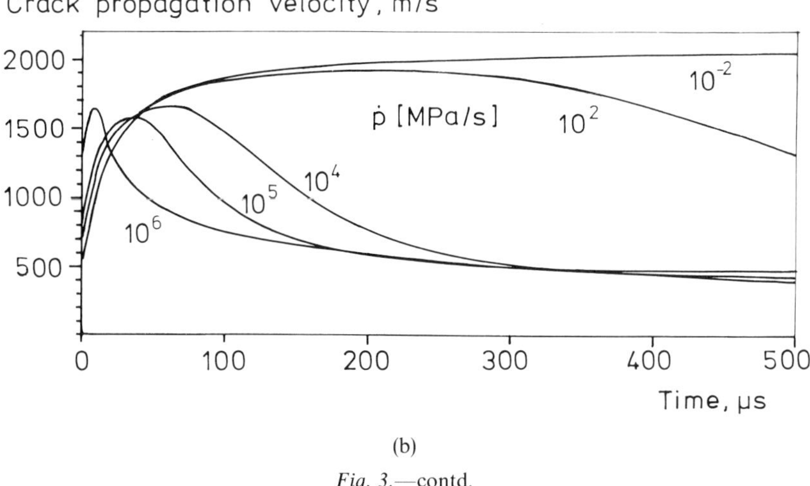

(b)

Fig. 3.—contd.

rate on the various energy components showed that the kinetic and deformation energy, normalized by the work, increased significantly with increasing loading rate. It seems that the rate of energy supply can become too high to be absorbed in the fracture process, resulting in an equilibrium in which a major part of the energy supplied is stored around the crack tip as kinetic and deformation energy. This should mean that the stress distribution around the crack tip changes and the stress intensity factor decreases with increasing loading rate due to the contribution of the inertia to the force equilibrium. This is the same result as Curbach [10] obtains from his finite element calculation on a notched specimen. It may be noted also that there is support for the foregoing results from the damage model of Chen [21].

Finite geometry affects the kinetic energy terms through the velocity field during crack extension (see Appendix.) In (5) the functions $AA(t)$ and $BB(t)$ increase with decreasing intermediate distance resulting in an equilibrium at a lower crack velocity.

It should be noted that the results of Fig. 4b show that the mean crack tip velocity increases in the fictitious fracture plane with increasing loading rates. This feature is consistent with experimental observations for macro cracking in [22] and [10].

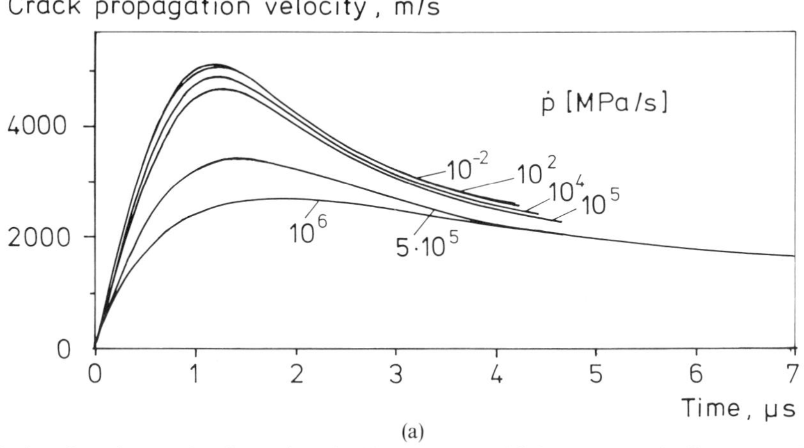

(a)

Fig. 4. The velocity of crack extension for various loading rates $\dot{p}$, and finite geometry ($a_0/b = 0.4$, see Fig. 1). $\dot{p} = 10^{-2}$, 10^2, 10^4, 10^5, $5\cdot10^5$, 10^6 MPa/s. (a) No adjustments for inertia effects. (b) Adjustments for inertia effects.

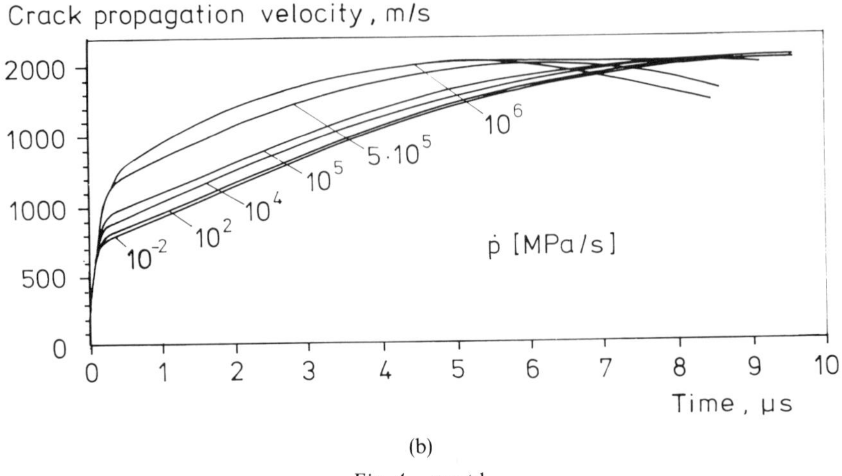

(b)

Fig. 4.—contd.

In the following section the crack extension model will be applied to the schematic failure plane for concrete.

4. Comparison of model prediction with test results

To determine the effect of the loading rate on the failure process and the tensile strength, crack propagation for various loading rates is compared with the process under static loading. Because no micro cracking, crack arrest or branching are modelled, no absolute values for the crack velocity or strength can be obtained without reference to the values under static loading. The rate effect on the strength follows from relating the processes modelled under dynamic and static loading [19].

To show how the effect of the model and loading rate on the crack extension is reflected in the dynamic tensile strength, first the fracture energy will be kept constant. Figure 5 shows the results for a high (*A*) and a low (*B*) quality concrete together with some experimental data. The difference in quality is characterized by a factor two in compressive strength. The lines *A* and *B* in Fig. 5 show that the model predicts the strength increase for high loading rates very well and also the more sensitive response for lower concrete quality.

From this result it must be concluded that the steep strength increase for high loading rates is not caused by an increase in fracture energy due to a different fracture path or multiple cracking, but is caused by a changing stress and energy distribution in the regions around the crack tip. From this result it emerges also that the increase at lower loading rates is caused by increasing energy demand, which has to be included in the model. Zielinski [23] studied the different fracture planes and multiple cracking under dynamic loading. His work has been used to incorporate these effects on the geometry of the fictitious fracture plane and the stress field. The results are given in the lines *A'* and *B'* of Fig. 5.

With this approach the predicted dynamic strength corresponds well with the experimental data for rates from 10^{-2} to 10^6 MPa/s. The effect of concrete quality is also represented well for the lower rates.

The phenomena described by the model are well known. For instance, the crack and damage extension at high strain rates led to the cube root relation of Kipp (3); recently Curbach applied

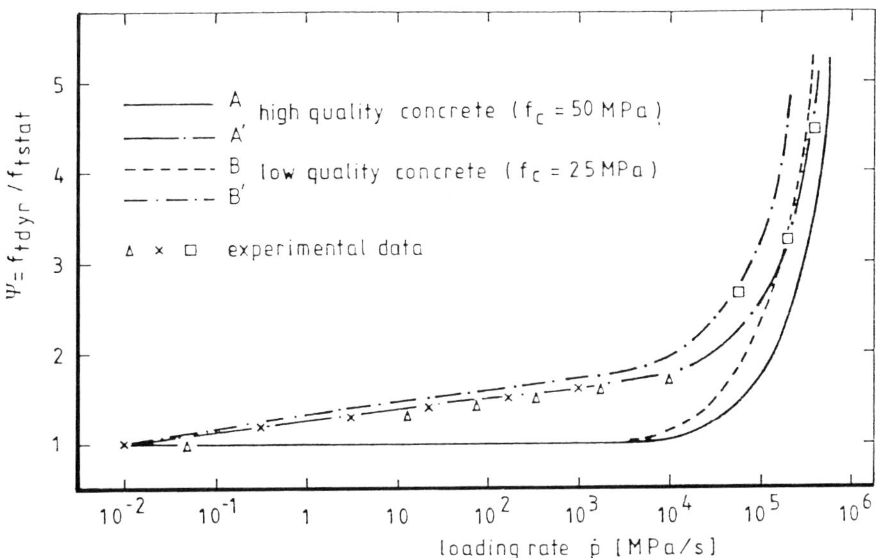

Fig. 5. Relative tensile strength $\Psi(=f_{td}/f_{ts})$ as a function of the loading rate $\dot{p}$. Experimental results from Zielinski [23], Takeda [24], and Birkimer on medium and low strength concrete.
Line *A*: high quality concrete; constant fracture energy.
Line *A'*: high quality concrete; fracture energy depends on *p*.
Line *B*: low quality concrete; constant fracture energy.
Line *B'*: low quality concrete; fracture energy depends on *p*.

the experimentally determined maximum velocity for macro cracks to explain the steep strength increase in the high rate region. The main inertia effect on material response, i.e. the limiting damage extension rate was also applied in the first version of the current model [25]. The newer work however reflects a more basic approach, using the conservation of energy; it led to an expression for damage extension process for all loading rates. Thus, the model offers the opportunity to examine the influence of concrete composition and material properties on the fracture process and, in particular, impact on dynamic strength.

5. Concluding remarks

It can be concluded that the moderate and steep increase of the tensile strength between low and high loading rates are caused by two different phenomena. For loading rates up to about 10 GPa/s the strength increase is caused by an increasing energy demand to form the final fracture plane. Aggregate fracture and multiple fracture occurs. The mechanism of the extension of the existing damage does not change for these loading rates and the dependency can be expressed as a power function with one constant coefficient (see Eqn. (2) of [6]). The steep increase beyond 10 GPa/s is caused by a mechanical response of the materials around the crack tip in which the inertia effects become dominant. The distribution of the energy supplied changes and K_{ID} decreases, thus causing a decreasing crack propagation velocity and a corresponding increase in strength.

The crack extension in the fictitious fracture plane showed that the crack velocity in linear elastic material is limited to the Rayleigh wave velocity but also that the terminal crack velocity decreases at higher loading rates to about $0.25\,C_r$. Comparison of this result with the observed crack velocities in real concrete is not justified because these are related to macro cracking in the post-peak fracture process in the non-linear and non-elastic material. However, the feature of decreasing terminal velocities at higher loading rates, and so increasing strength, remains valid.

In (1) the fracture energy G_F is coupled to the tensile strength under the condition that no multiple cracking occurs in the measured length of the fracture zone. The aspect of multiple cracking and the determination of G_F from dynamic tests is also discussed in [26]. From the different mechanisms indicated by the model, it emerges that (1) can only be valid for moderate loading rates, because for higher loading rates the rate dependency of the strength is not coupled to the dissipated fracture energy.

Finally, in spite of the assumptions and simplifications, the model describes well the rate dependency of the uniaxial tensile strength for all loading rates and reflects also the rate sensitivity for different concrete qualities. Apparently the most important aspects of the material response of concrete under dynamic loading, but before ultimate load, can be characterized by the extension of the existing damage in the fictitious fracture plane.

Appendix: The energy terms

The terms are given by:
- The work $W(t)$ is determined by the stress component, normal to the surface A and the displacement of the material at the envelope.

$$W(t) = \iint_A (\sigma_n,\, \mathrm{d}u)\, \mathrm{d}A,$$

where σ_n is the component of the stress vector normal to the surface A and $\mathrm{d}u$ is a component of the differential of the displacement vector.

$$W(t) = 2\frac{(1+v)}{E} \cdot CV_t \int_t \left[p\dot{p}af^2 + \frac{p^2a}{2}(f^2 + fg) \right] \mathrm{d}\tau,$$

with $v =$ Poisson's ratio, $p =$ loading, $f =$ geometry function, $E =$ Young's modulus, $\dot{p} =$ loading rate, $g =$ geometry function.

The deformation energy $V(t)$ in A is determined by the stress and deformation fields, and follows from volume integration of the vector product of both components.

$$V(t) = 0.5 \int_V \sigma \cdot \varepsilon \, \mathrm{d}V,$$

where ε is the strain vector and V the volume inside A.

$$V(t) = \frac{(1+v)}{E} \cdot CV_t \cdot p^2 af^2.$$

Also the kinetic energy is stored inside A and can be determined when the velocity field is known.

$$T(t) = 0.5\rho \int_V (\dot{u})^2 \, \mathrm{d}V,$$

with ρ the density of the material and u the velocity.

$$T(t) = \frac{(1+v)^2}{E^2} \cdot \rho \cdot CT_t \left[a\dot{p}^2 f^2 + p\dot{p}\dot{a}(f^2 + fg) + \frac{p^2}{4a}\dot{a}^2(f+g)^2 \right].$$

– The fracture energy $D(t)$ is proportional to the specific surface energy γ and the area of the created fracture planes. For penny-shaped cracks $D(t)$ is given by

$$D(t) = 2 \cdot \gamma \cdot \pi \cdot a^2.$$

In these expressions the coefficients CV_t and CT_t are the result of the surface and volume integrals and given by

$$CV_t = CV(t) = (5aR + 2R^2) - v(8aR + 4R^2),$$

$$CT_t = CT(t) = \left(\frac{38}{3} - 32v + \frac{64}{3}v^2 \right)aR^3 - R^4(3 - 4v),$$

with $R = \lambda$ times characteristic length.

References

1. R. John and S.P. Shah, Mixed mode fracture of concrete subjected to impact loading, NSF-ACBM Report, Northwestern University, Evanston, January 1989, 1–18.
2. A. Hillerborg, *International Journal of Cement Composites* 2 (1980) 177–184.
3. D.A. Hordijk and H.W. Reinhardt, in *Brittle Matrix Composites* 2, A.M. Brandt and I.H. Marshall (eds.), Elsevier Applied Science Series, London (1988) 486–495.
4. H.A. Kormeling and H.W. Reinhardt, *International Journal of Cement Composites and Lightweight Concrete* 9 (1987) 197–204.
5. A.S. Krausz and K. Krausz, *Fracture Kinetics of Crack Growth*, Kluwer Academic Publishers, Dordrecht (1988).
6. H. Mihashi and F.H. Wittmann, *Heron* 25 (1980) 1–54.
7. M.E. Kipp, D.E. Grady and E.P. Chen, *International Journal of Fracture* 17 (1980) 471–478.
8. S.P. Shah and R. John, in *Fracture Toughness and Fracture Energy of Concrete*, F.H. Wittmann (ed.), Elsevier, Amsterdam (1986) 453–465.
9. D. Muria Vila and P. Hamelin, in *Proceedings 1st RILEM Congress* 2, Chapman & Hall, London (1987) 725–732.
10. M. Curbach, Festigkeitssteigerung von Beton bei hohen Belastungsgeschwindigkeiten, Dr.-Ing. dissertation, Karlsruhe, Germany, 1987.
11. J. Weerheijm, Schematisation of the fracture process in concrete under compressive and tensile loading, Research report TNO-PML No. 1987–16, Delft (1987) in Dutch.
12. L.B. Freund, *Journal of the Mechanics and Physics of Solids* 21 (1973) 47–61.
13. K.B. Broberg, *Arkiv fur Fysik* 18 (1960) 159–192.
14. C. Atkinson and J.D. Eshelby, *International Journal of Fracture Mechanics* 4 (1968) 3–8.
15. J.D. Achenbach and L.M. Brock, in *Dynamic Crack Propagation*, G.H. Sih (ed.), Noordhoff, Leyden (1972).

16. H. Nozaki et al., *International Journal of Solids and Structures* 22 (1986) 1137–1147.
17. N.F. Mott, *Engineering* 165 (1948) 16–18.
18. J.P. Berry, *Journal of the Mechanics and Physics of Solids* 8 (1960) 194–206.
19. J. Weerheijm, Properties of Concrete Under Dynamic Loading 3. Fracture Model for Brittle Materials, PML 1990–9, Rijswijk (1990).
20. I.N. Sneddon, in *Mechanics of Fracture 1. Methods of Analysis and Solutions of Crack Problems*, G.C. Sih (ed.), Noordhoff, Leyden (1973).
21. E.P. Chen, in *Proceedings Symposium on Cement-Based Composites: Strain Rate Effects on Fracture*, Materials Research Society Symposium Proceedings 64 (1986) 63–77.
22. R. John and S.P. Shah, A fracture mechanics model to predict the rate sensitivity of mode I fracture of concrete, Northwestern University, Evanston, November 1986.
23. A.J. Zielinski, *Fracture of Concrete and Mortar Under Uniaxial Impact Tensile Loading*, Delft University Press, Delft (1982).
24. J. Takeda and H. Tachikawa, in *Proceedings International Conference on Mechanical Behaviour of Materials*, Kyoto IV (1971) 267–277.
25. J. Weerheijm and W. Karthaus, in *Proceedings Symposium on The Interaction of Non-Nuclear Munitions with Structures*, Panama City Beach (1985).
26. J. Weerheijm and H.W. Reinhardt, in *First International Conference on Structures Under Shock and Impact*, Cambridge, Massachusetts (1989).

PART 2.   NUMERICAL FRACTURE ANALYSIS,
DESIGN CODES AND APPLICATIONS

International Journal of Fracture **51**: 45–59, 1991.
Z.P. Bažant (ed.), *Current Trends in Concrete Fracture Research.*
© 1991 *Kluwer Academic Publishers. Printed in the Netherlands.*

Smeared and discrete representations of localized fracture

JAN G. ROTS
Delft University of Technology, Faculty of Civil Engineering, P.O. Box 5048, 2600 GA Delft, The Netherlands

Received 11 June 1990; accepted 11 November 1990

Abstract. The possibilities of smeared and discrete crack concepts for simulating localized fracture in softening materials are investigated. First, comparisons are made between fixed, multi-directional and rotating smeared cracks, whereby the crack orientation is kept constant, updated in a stepwise manner or updated continuously, respectively. Next, the smeared approaches are compared to analyses on systems of predefined potential discrete cracks. Results indicate that (1) fixed smeared cracks may produce overstiff behavior while rotating smeared cracks do not, (2) smeared cracks may give rise to stress-locking while discrete cracks do not. Examples are shown for concrete and masonry.

1. Introduction

Tensile fracture in matrix-aggregate composites like concrete involves progressive micro-cracking, tortuous debonding and other processes of internal damage. These processes eventually coalesce into a geometrical discontinuity that separates the material. Such a discontinuity is called a crack.

The discrete crack concept is the approach that reflects the final damaged state most closely. It models the crack directly via a displacement-discontinuity in an interface element that separates two solid elements. Unfortunately, the approach does not fit the nature of the finite element displacement method and it is computationally more convenient to employ a smeared crack concept. A smeared crack concept imagines the cracked solid to be a continuum and permits a description in terms of stress-strain relations. However, here the converse drawback occurs, since the underlying assumption of displacement continuity conflicts with the realism of a *dis*continuity.

To date, there is no consensus on the question of which type of approach should be preferred. A further complication is that the smeared crack concept itself already offers a variety of possibilities, ranging from fixed single to fixed multi-directional and rotating crack approaches. Here, the distinction lies in the orientation of the crack, which is either kept constant, updated in a stepwise manner or updated continuously. The differences emerge when the principal stress rotates after cracking, as is typical of general fracture analysis.

It is the purpose of this paper to investigate the merits and demerits of the various approaches. Recently, a few similar comparative studies have been presented, but these were mainly concerned with single-element problems of distributed fracture (e.g. Balakrishnan and Murray [1], Willam et al. [2], Barzegar [3], Chrisfield and Wills [4]). In the present paper the focus is placed on localized fracture in element assemblies. Examples are shown for concrete and masonry.

2. Discrete crack concept

The discrete crack concept models a crack by means of a separation between element edges (e.g. Ngo and Scordelis [5]). The approach implies a continuous change in nodal connectivity and constrains the crack to follow a path along the element edges. For general purposes this is not exactly attractive, but a number of special purposes exist for which the drawbacks can be circumvented. These relate to cases whereby the crack orientation does not constitute the prime subject of interest. This for example occurs if we employ smeared approaches as a predictor of the crack path, whereafter a corrector analysis is made on a mesh in which a potential discrete crack is predefined according to the smeared prediction. For comparative studies this is an effective strategy. Another example is engineering problems whereby a mechanism of discrete cracks is preimagined in a fashion similar to yield line mechanisms.

For such problems one may predefine interface elements in the original mesh, keeping the topology preserved. The initial stiffness of the elements is assigned a large dummy value in order to simulate the uncracked state with rigid connection between overlapping nodes. Upon violating a condition of crack initiation, the element stiffness is changed and a constitutive model for discrete cracks is mobilized. Such a model links the tractions t^{cr} to the relative displacements u^{cr} across the crack via C^{cr} which represents phenomena like softening:

$$\Delta t = C^{cr}\Delta u^{cr}. \tag{1}$$

Caution should be exercised in selecting the type of interface element. A distinction can be made between lumped interface elements [5] which evaluate the tractions and displacements at isolated node-sets, and continuous interface elements (Goodman et al. [6]) which smooth the behavior along an interpolated field. It has sometimes been suggested that the latter class of elements is superior. However, Rots [7] and, independently a number of researchers from geomechanics (e.g. Hohberg and Bachmann [8]) have found that this is not generally true. This relates to the question of how large the dummy stiffness can and should be made. Ideally, it should be made extremely large to keep the initial dummy separation negligible. With the continuous interface elements, such high stiffness values turned out to produce significant flutter in the traction profiles, especially for Gaussian integration schemes, whereas with the lumped interface elements the results for increasing dummy stiffness correctly converged towards the rigid-connection solution. For this reason, only the lumped elements are employed in this paper.

3. Fixed single smeared crack concept

A consistent way to set-up a fixed smeared crack concept is to decompose the global strain increment $\Delta\varepsilon$ into a part $\Delta\varepsilon^{cr}$ of the crack and a part $\Delta\varepsilon^{co}$ of the solid material between the cracks (de Borst and Nauta [9], Rots et al. [10]):

$$\Delta\varepsilon = \Delta\varepsilon^{cr} + \Delta\varepsilon^{co}. \tag{2}$$

It will be shown that this opens up the possibility to put fixed single, fixed multi-directional and rotating smeared cracks into the same picture. Attention is restricted to a two-dimensional configuration, but this is not essential.

The strain vectors in (2) relate to the global x,y coordinate axes. In addition, let us set up a local n,t-coordinate system aligned with the crack. In this system, the local crack strain vector is defined as Δe^{cr}

$$\Delta e^{cr} = [\Delta\varepsilon_{nn}^{cr}\,\Delta\gamma_{nt}^{cr}]^T, \tag{3}$$

where ε_{nn}^{cr} is the mode I crack normal strain and γ_{nt}^{cr} is the mode II crack shear strain, and ^{T} denotes a transpose. The relation between local and global crack strains reads

$$\Delta\varepsilon^{cr} = N\Delta e^{cr}, \tag{4}$$

with N being a transformation matrix that reflects the orientation of the crack. This matrix is assumed to be fixed upon crack initiation, so that we deal with a fixed crack concept. In the local coordinate system, we furthermore define a vector Δt^{cr} or incremental tractions across the crack:

$$\Delta t^{cr} = [\Delta t_n^{cr}\,\Delta t_t^{cr}]^T, \tag{5}$$

in which Δt_n^{cr} is the mode I normal traction and Δt_t^{cr} is the mode II shear traction increment. The relation between the global stress increment $\Delta\sigma$ and the local traction increment can be derived to be

$$\Delta t^{cr} = N^T\Delta\sigma. \tag{6}$$

To complete the system of equations, a constitutive model is needed for the concrete and a traction-strain relation for the smeared cracks. For the concrete between the cracks, the following relationship is assumed:

$$\Delta\sigma = D^{co}\Delta\varepsilon^{co}, \tag{7}$$

with D^{co} containing the instantaneous moduli of the concrete. For the crack, the smeared crack analogy of (1) is employed:

$$\Delta t^{cr} = D^{cr}\Delta e^{cr}, \tag{8}$$

with D^{cr} a 2×2 crack matrix incorporating the mode I, mode II and, eventually, mixed-mode properties of the crack. Using (2)–(8), the overall stress-strain law for the cracked concrete can be developed:

$$D^{co}\Delta\sigma = [D^{co} - N[D^{cr} + N^TD^{co}N]^{-1}N^TD^{co}]\Delta\varepsilon. \tag{9}$$

4. Fixed multi-directional smeared crack concept

The strategy of strain-decomposition allows for a sub-decomposition of the crack strain into the separate contributions of a number of multi-directional cracks that simultaneously occur at

48   *J.G. Rots*

a sampling point, i.e.

$$\Delta \varepsilon^{cr} = \Delta \varepsilon_1^{cr} + \Delta \varepsilon_2^{cr} \cdots,$$

(10)

where $\Delta \varepsilon_1^{cr}$ is the global crack strain increment owing to a primary crack, $\Delta \varepsilon_2^{cr}$ is the global crack strain increment owing to a secondary crack and so on. Each of these (fixed) cracks is assigned its own local crack strain vector e_i^{cr}, its own traction vector t_i^{cr}, its own fixed transformation matrix N_i and its own traction-strain matrix D_i^{cr}. By assembling these individual vectors and matrices, the overall stress-strain relation for the multiply cracked solid can be set up analogously to the relation for a singly cracked solid [9].

The multi-directional crack concept is applicable to conditions where the fracture starts in tension and subsequently proceeds in tension-shear. This behavior generally implies that the axes of principal stress rotate after crack formation. For such cases, the use of fixed single cracks leads to an increasing discrepancy between the axes of principal stress and the fixed crack axes. The fixed multi-directional crack concept provides an alternative. Whenever the angle of inclination between the existing crack(s) and the current direction of principal stress exceeds the value of a certain threshold angle α, a new crack can be initiated. In this way, we end up with a system of non-orthogonal cracks, as pioneered in [9].

5. Rotating smeared crack concept

A prominent feature of the fixed single and fixed multi-directional crack concepts is the inclusion of an *explicit* shear term for the fixed plane(s). Unfortunately, the identity 'shear' does not provide much insight into the behavior of structures. Designers as well as researchers prefer to think in terms of 'principal stress', the axes of which continuously rotate after cracking. This poses a need to monitor stress-strain relations in the rotating principal coordinate system, instead of in the fixed crack system.

The practical use of such rotating principal stress-strain relations requires principal stress and strain to be coaxial. It has been shown that this coaxiality can be achieved only via an *implicit* shear term in the rotating principal 1, 2 reference frame [11], [2]:

$$G_{12} = \frac{(\sigma_{11} - \sigma_{22})}{2(\varepsilon_{11} - \varepsilon_{22})}.$$

(11)

The author has implemented this so-called rotating crack concept as the limiting case of the fixed multi-directional crack concept [7]. A zero threshold angle is employed, such that a new crack of slightly rotated orientation arises in each load step. The previous cracks are erased from memory as soon as the new crack arises, and the shear modulus across the new crack is taken such that (11) is satisfied. An advantage of conceiving the rotating crack approach as the limiting case of the fixed multi-directional crack approach is that it maintains decomposition of total strain into concrete strain and crack strain. This permits the concrete material law (e.g. elasticity) to be satisfied exactly. With other versions of the rotating crack concept this is not necessarily true as these models generally lose Poisson's effect after cracking [1, 4, 12, 2].

All formulations in Sections 3–5 have been presented in incremental form. This requires provisions to handle nonlinear fracture functions (e.g. a nonlinear tensile softening curve, or a non-constant shear term) and state changes owing to initiation, closing and re-opening of cracks. To this end, a combination of a forward-Euler inner iteration loop with sub-division of the strain path has been employed.

6. Elastic-softening constitutive model

For tension and tension-shear dominated fracture an adequate model is constructed by assuming elasticity for the concrete and softening for the crack. Consequently, D^{co} of (7) is the elasticity matrix. C^{cr} of (1) for discrete cracks and D^{cr} of (8) for smeared cracks are assumed to be of the form:

$$C^{cr} = \begin{bmatrix} C^I & 0 \\ 0 & C^{II} \end{bmatrix}, \tag{12}$$

$$D^{cr} = \begin{bmatrix} D^I & 0 \\ 0 & D^{II} \end{bmatrix}. \tag{13}$$

The assumption of zero off-diagonal terms implies that *direct* shear-normal coupling across a fixed plane has been ignored. It will be demonstrated that this is justified since a similar phenomenon can be obtained *indirectly*, by allowing for crack rotation.

The mode I stiffness moduli C^I and D^I are set by three tensile softening parameters: the tensile strength f_{ct}, the fracture energy G_f and the shape of the softening diagram. For a discrete crack, this relation can be worked out exactly [13], whereas for a smeared crack a fourth factor enters the scene, viz. the crack band width h, which is related to the particular finite element configuration [14]. The tensile-softening diagram is assumed to be a fixed material property. A recent scrutiny of the direct tension problem, [15, 16], has shown that the concave shape of the diagram is essential, as shown in Fig. 1.

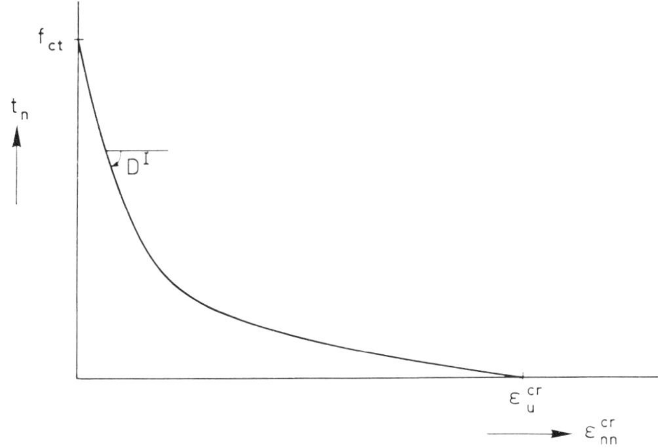

Fig. 1. Concave mode I tensile softening relation between crack normal traction and crack normal strain [15].

In this paper, the mode II stiffness moduli C^{II} and D^{II} have not been set-up in the spirit of mode II or mixed-mode fracture. This is because there is no consensus on the question of whether such phenomena exist for heterogeneous materials like concrete. Aside from cases of extremely high mode II or mode III intensity – which reach beyond the scope of the present paper – the relevant experiments reveal that curved fractures are predominantly of the mode I type [17–20]. Consequently, any terminology of mode II in this paper relates solely to stress rotation at the local integration point level. For the discrete crack concept, stress rotation occurs in the elastic elements at either side of the discontinuity. Hence, any mode II phenomena can be constrained from the analysis, so that C^{II} has been set equal to zero after discrete cracking. For the smeared crack concept, D^{II} is highly essential as it controls the stress rotation in cracked elements. It is illustrative to relate this modulus to the classical shear retention factor β [21]. This can be done by considering that the strain-decomposition implies

$$\frac{1}{\beta G} = \frac{1}{G} + \frac{1}{D^{II}} \rightarrow D^{II} = \frac{\beta}{1 - \beta} G, \tag{14}$$

in which G is the elastic shear modulus. In this paper, numerical experiments are undertaken on three formulations:

- $\beta = 0$, i.e. $D^{II} = 0$. This implies zero shear transfer and no stress rotation.
- $\beta = 0.05$, i.e. $D^{II} \approx 0.053G$. This is the traditional method of some constant shear retention factor.

- $$\beta = \left[1 - \frac{\varepsilon_{nn}^{cr}}{\varepsilon_u^{cr}} \right]^p, \tag{15}$$

  with p some constant and ε_u^{cr} being the ultimate strain of the softening diagram. This degradation of β from 1 (full retainment of elastic shear) to zero, corresponds to a degradation of D^{II} from infinite (full interlock) to zero. This relation involves decreases of the shear resistance with increasing crack opening. It is objective with respect to mesh refinement, since ε_u^{cr} is adapted to the crack band width.

For the extension towards multi-directional cracks, it is essential that the energy consumed in previous defects is subtracted from the energy available for the current defect, and that the variable shear retention factor is made a function of the principal crack strain ε_{11}^{cr}, representing the damage in all cracks, instead of ε_{nn}^{cr} for a single crack. The further extension towards rotating cracks lies in a zero threshold angle α and in the coaxiality-enforcing shear term of (11). βG along the 'currently-active crack' must be taken in accordance with this shear term.

7. Tension-shear model problem

To illustrate the gradual transition from fixed single, fixed multi-directional to rotating cracks, reference is made to a tension-shear model problem. It concerns an elastic-softening continuum of unit dimensions with $E = 10000 \, \text{N/mm}^2$, $v = 0.2$, $f_{ct} = 1.0 \, \text{N/mm}^2$ and a linear strain-softening diagram with an ultimate strain ε_u^{cr} that equals three times the elastic strain at peak-strength, i.e. $\varepsilon_u^{cr} = 0.0003$, corresponding to $G_f = 0.15 \, \text{J/m}^2$ over unit crack band width $h = 1.0 \, \text{mm}$. Initially, the continuum is subjected to tensile straining in the x-direction accom-

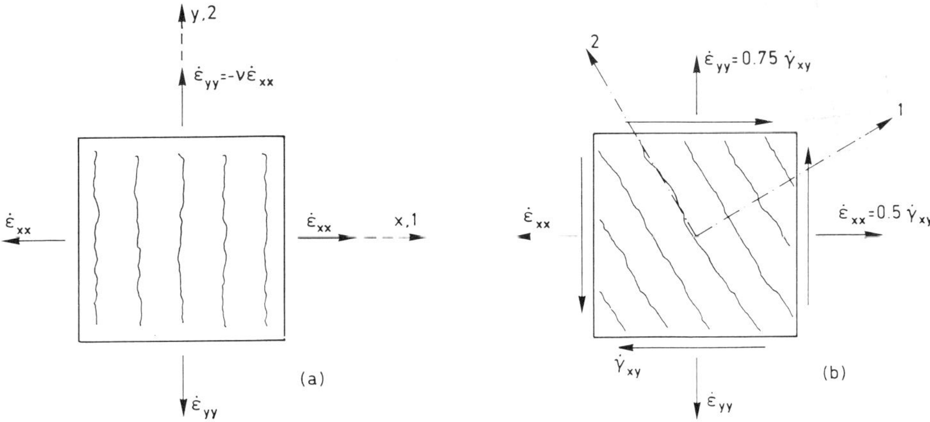

Fig. 2. Lay-out of tension-shear model problem, (a) tension up to cracking; (b) biaxial tension with shear beyond cracking.

panied by lateral Poisson contraction in the y-direction, i.e. $\Delta\varepsilon_{xx}:\Delta\varepsilon_{yy}:\Delta\gamma_{xy} = 1:-\nu:0$. Immediately after cracking, a switch is made to combined biaxial tension and shear according to $\Delta\varepsilon_{xx}:\Delta\varepsilon_{yy}:\Delta\gamma_{xy} = 0.5:0.75:1$, involving the axes of principal strain to continuously rotate after crack formation. This scheme is shown in Fig. 2 and was inspired by recent work of William et al. [2].

The focus is placed on a variation of the threshold angle α, while the mode II relation across the currently-active crack is kept fixed in the form of a quadratic degradation of β according to (15) with $p = 2$ (different crack shear relations do not affect the basic conclusions). The adopted values of α range from 0, 7.5, 15, 30, 45 up to 90 degrees. The extreme of $\alpha = 0°$ corresponds to a continuously rotating crack, not maintaining coaxiality however, and the other extreme of $\alpha = 90°$ corresponds to a fixed single-crack. Note that the fixed single-crack result for this homogeneous test problem can be obtained by using both a smeared and a discrete crack (a discrete crack is by definition fixed single). In addition, the problem was analyzed using the coaxial rotating crack model, i.e. for threshold angle $\alpha = 0°$ and β enforcing coaxiality according to (11). Summarizing, the following cases have been studied:

- Fixed single-crack (non-coaxial):
  threshold angle $\alpha = 90°$;
  variable shear retention factor β according to (15) with $p = 2$.
- Multi-directional cracks (non-coaxial):
  threshold angle $\alpha = 0, 7.5, 15, 30$ and $45°$;
  variable shear retention factor β according to (15) with $p = 2$.
- Rotating cracks (coaxial):
  threshold angle $\alpha = 0°$;
  variable shear retention factor β according to (11).

The nominal $\tau_{xy} - \gamma_{xy}$ shear response predicted is shown in Fig. 3. The two computations for a zero threshold angle provide the most flexible response. In addition, these two computations show curves that are nicely smooth as a result of the continuously changing orientation of the crack. Upon increasing the threshold angle, the response becomes less flexible and we observe increasing discontinuity as a result of the interval adaptation of the currently-active crack

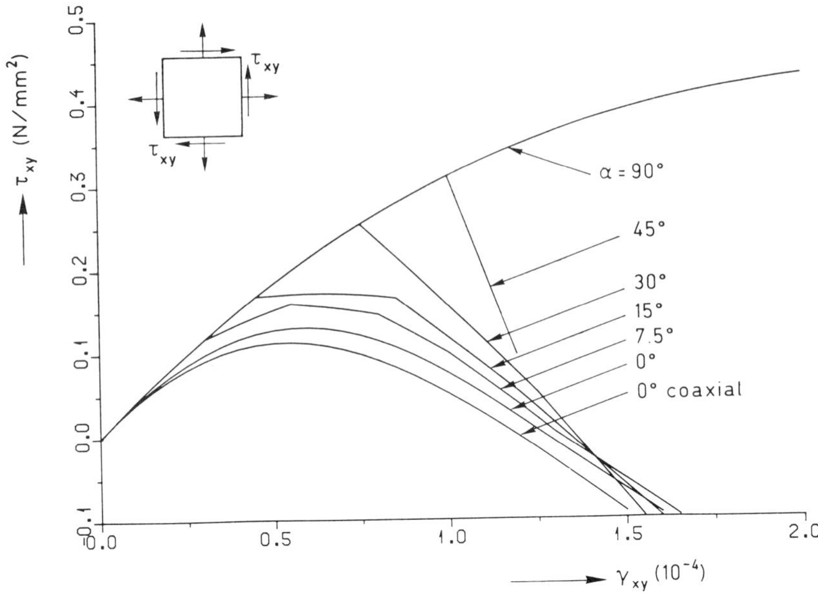

Fig. 3. Shear stress-strain response of model problem in x, y direction, for different threshold angles α between multi-directional cracks.

orientation to the continuous rotation of principal stress. For threshold angles beyond 45° the initiation of inclined cracks is postponed even further and the solution approaches the fixed single-crack solution for $\alpha = 90°$, which is extremely stiff.

The observation of stiff behavior for fixed-single cracks corresponds to previous findings (e.g. [1, 4, 22, 2]). The present results indicate that a crack-strain decreasing shear retention function does not provide a remedy. A remedy is provided by reducing the threshold angle. Figure 3 shows that the results for threshold angles less than 90° invariably display shear softening along the x,y-plane of initial cracking. This shear softening occurs implicitly, as a consequence of the crack rotation. In a similar way, it can be shown that implicit shear-normal coupling occurs as a consequence of crack rotation [7].

Figure 4 shows the principal tensile stress σ_{11} versus principal tensile strain ε_{11}. The fixed-single crack curve ($\alpha = 90°$) is very illustrative as it shows σ_{11} decreases only slightly beyond crack initiation, whereafter it re-starts to increase and amply exceeds the tensile strength. On decreasing the threshold angle, inclined cracks arise and the tensile strength is correctly kept under control, which is reflected by the truncation of the fixed single crack curve. The curve for coaxial rotating cracks replicates the input softeing diagram, since it monitors the softening explicitly in the rotating principal system.

It is concluded that crack rotation is an effective strategy for achieving a flexible response that correctly keeps the principal tensile stress under control.

8. Single-notched shear beam

To investigate whether the above conclusion also holds for structural fracture in element assemblies, we consider a single-notched shear beam which falls in curved mode I fracture [17].

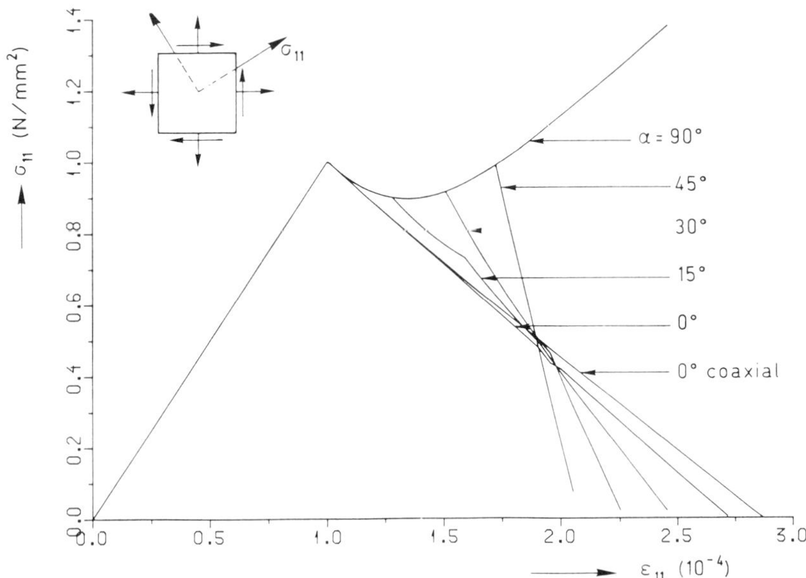

Fig. 4. Principal tensile stress-strain response of model problem in 1, 2 direction, for different threshold angles α between multi-directional cracks.

The beam has been analysed before (e.g. [17, 23, 22, 24]), but none of the smeared crack solutions presented so far has displayed genuine separation and softening down to zero. Mostly a too stiff response was found, which may be related to the issues discussed in the preceding section.

The mesh is shown in Fig. 5 and consists of three-node triangles in a cross-diagonal pattern. The steel beam ACB has not been included in the mesh. Instead, the loading has been applied at the points A and B, while controlling the notch tip opening-displacement in order to capture the snap-back involved [23]. The parameters were taken as: $E = 24800\,\text{N/mm}^2$, $v = 0.18$, $f_{ct} = 2.8\,\text{N/mm}^2$, $G_f = 100\,\text{J/m}^2$, concave softening (Fig. 1) and $h = 12\,\text{mm}$.

In order not to overinform the reader, attention is confined to the extreme cases of fixed single and/or orthogonal cracks (i.e. a high threshold angle) and coaxial rotating cracks. Five results will be discussed:

- fixed cracks, $\alpha = 60°$, $\beta \approx 0$;
- fixed cracks, $\alpha = 60°$, $\beta = 0.05$;
- fixed cracks, $\alpha = 60°$, variable β according to (15) with $p = 2$;
- coaxial rotating cracks, i.e. $\alpha = 0$, β enforcing coaxiality according to (11);
- predefined discrete crack, zero shear stiffness and zero shear tractions.

Its location was adapted to the average of the smeared crack prediction. Possible compression nonlinearity under the roller bearing platen has not been included in the analyses. For this reason, quantitative comparisons with the experiment will not be drawn. Instead, we will concentrate on comparisons between the various smeared results and the discrete result, whereby the tensile cracking parameters are kept constant.

The (fully converged) solutions are summarized in Fig. 6, giving the load F versus the displacement of the master loading point C, which is recalculated from the displacements of the minor loading points A and B on the assumption that the steel beam is infinitely stiff.

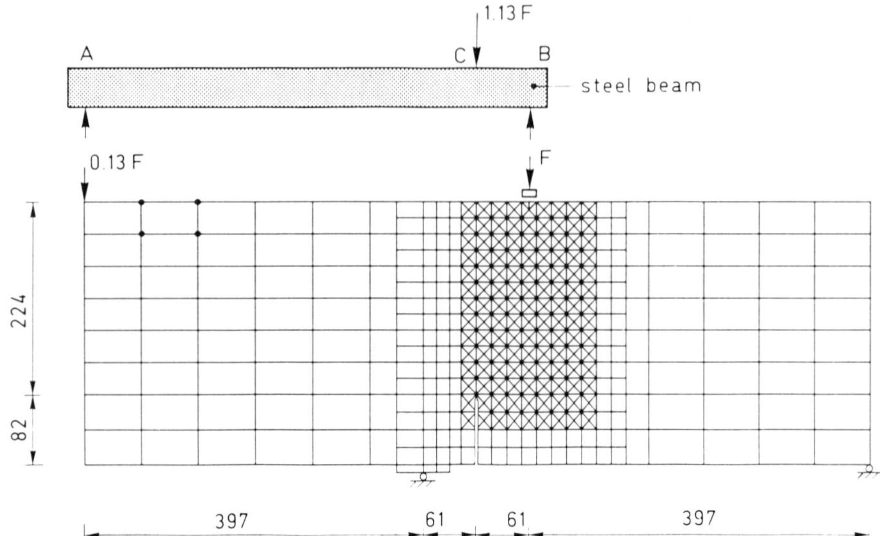

Fig. 5. Finite element idealization for smeared crack analysis of single-notched shear beam.

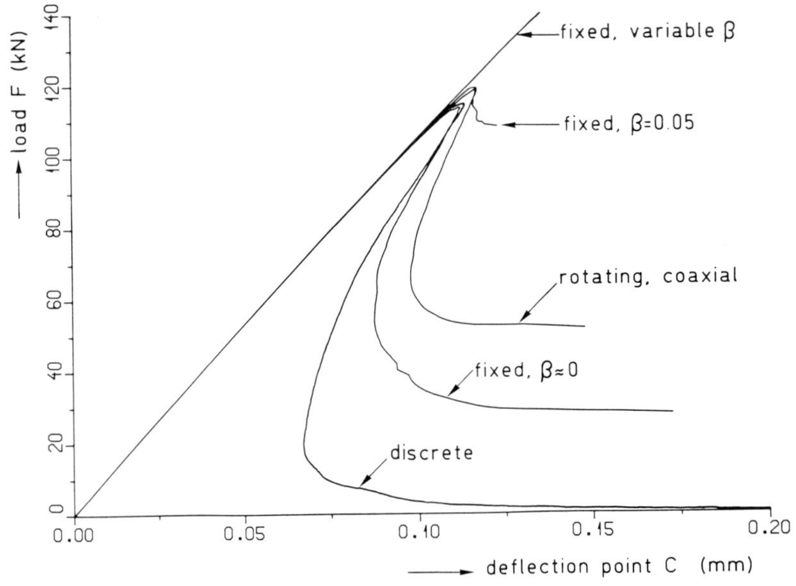

Fig. 6. Load *F* versus deflection point C of single-notched shear beam.

Let us first compare the various smeared crack results to one another. All smeared crack results are too stiff in the post-peak regime. Only fixed cracks with $\beta \approx 0$ and coaxial rotating cracks appear capable of producing a distinct softening response. Fixed cracks with significant shear retention fail in this respect. The latter statement already holds for a β value of 0.05, which is low compared to what is commonly used in engineering practice. The result for $\beta = 0.05$ shows far too little structural softening. For the variable shear retention factor that decreases with increasing crack strain, or any other high shear retention factor, the response becomes even worse as no softening is predicted at all.

The fact that fixed smeared cracks perform adequately only in conjunction with a zero or almost zero shear retention factor is surprising, yet explainable. Such a model in which crack shear tractions remain zero or almost-zero, implies the axes of principal stress to be fixed or practically fixed after crack formation. Stress rotation is then caused by the surrounding elastic elements, while any rebuild of principal tensile stress in the cracked elements is constrained to occur parallel to the first crack. Non-zero shear retention factors provide an additional possibility of rebuilding principal tensile stress, via rotation in an inclined direction. Whenever such an additional opportunity of stress rebuild is given, it is eagerly utilized. This provides a straightforward extension of the conclusion from the model problem. Again, crack rotation remedies the overstiff behavior, because the maximum principal tensile stress is correctly kept under control.

Secondly, let us compare the smeared crack results with the discrete crack result. It is striking that even the two best possible smeared crack results (fixed cracks with $\beta \approx 0$ and coaxial rotating cracks) do not produce softening down to zero. Instead, they produce an incorrect residual load plateau (Fig. 6). Softening down to zero could only be achieved via the computation on the discrete crack. To trace the causes, Fig. 7 gives an impression of the principal tensile stresses, both for a smeared crack result (coaxial rotating) and for the discrete crack result. Figure 8 gives the corresponding plots of the deformations. With the discrete crack analysis, we observe that the stressplot is undisturbed and that the stresses at either side of the separation correctly come down to zero, which is in agreement with the physical process. With the smeared crack analysis, spurious tensile stresses remain at either side of the separation. These stresses are locked-in because of the assumption of

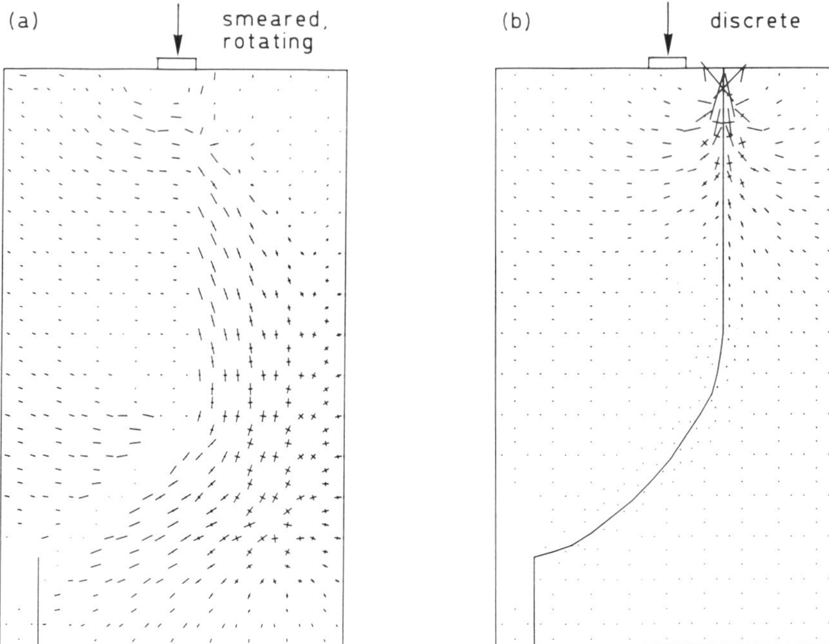

Fig. 7. Principal tensile stresses for single-notched shear beam at final stage. (a) Stress-locking for smeared cracks (coaxial, rotating); (b) Correct stress relief at either side of predefined discrete crack.

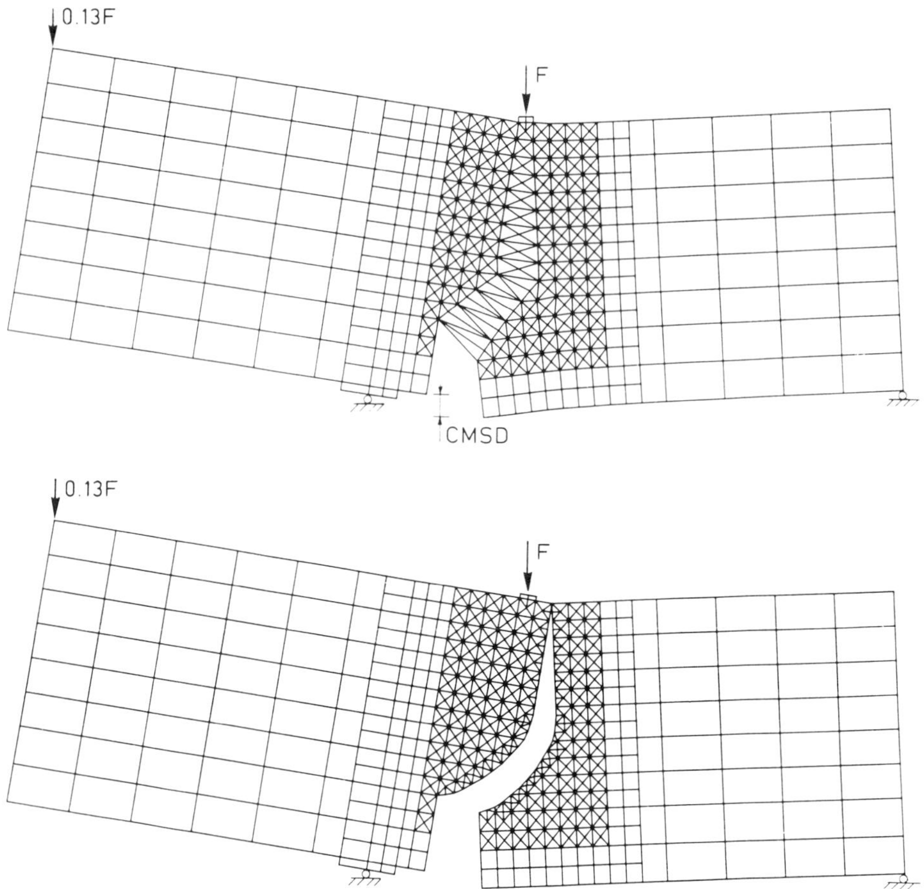

Fig. 8. Deformed meshes for single-notched shear beam at final stage. (a) Localization for smeared cracks (coaxial, rotating); (b) Genuine separation for predefined discrete crack.

displacement compatibility, as is inherent to smeared analyses. The locked-in stresses explain the too stiff response. For fixed smeared cracks with significant shear retention, these phenomena are even more pronounced because of the additional stress rebuild.

For details about the generality and the seriousness of stress-locking, the reader is referred to Rots [7, 25]. It is noted here that (1) the phenomenon was not found to disappear on mesh refinement, (2) the phenomenon occurred independent of the finite element types employed (linear and quadratic triangles and quadrilaterals), and (3) the phenomenon has only been demonstrated for local softening models.

The discrete crack analysis, which may serve as a valuable corrector after a smeared predictor, starts from displacement *incompatibility* and reveals genuine separation without stress-locking. This solution gives an excellent demonstration of the elastic-softening theory in that the energy supplied to the beam, which is represented by the integral of the master load $1.13F$ versus the master loading point displacement C in Fig. 6, precisely balances the fracture energy G_f times the surface area generated.

9. Masonry example

Fracture models for concrete have significant spin-off to similar materials like rock, brick, clay, mortar, ceramics and composites. This section considers an application to masonry. It is here that the method of potential discrete cracks is particularly attractive. The masonry is thereby modelled on a micro-level, i.e. the bricks and joints are treated separately by continuum elements and discontinuum (interface) elements respectively. Placing the interface elements in the horizontal joints, in the vertical joints and in the continuation of the vertical joints, the fracture can choose to propagate horizontally, vertically as well as in a zig-zag manner. Assuming linear-elastic behavior for the bricks and softening behavior for the cracks, the elastic-softening behavior of masonry walls can be simulated.

Figures 9 and 10 present a typical result. The geometrical and material properties have been taken from a wall tested by Ali and Page [26]. The softening was modeled at a linear curve with $G_f = 60\,\text{J/m}^2$ and the computation was performed under arc-length control. The load-displacement response of Fig. 9 is almost linear up to the peak load. Obviously, in this traject the effect of cracking is negligible. After peak load, both the load and the displacement suddenly decrease. This strong snap-back behavior is associated with the sudden propagation of a splitting type crack in the centre of the wall, as shown in Fig. 10.

For different load cases, e.g. horizontal loads in addition to vertical loads, it is not only the softening behavior but also the shear friction capacity of the joint-brick interface that affects the response. Further research is required in order to assess the proper interface properties for these cases.

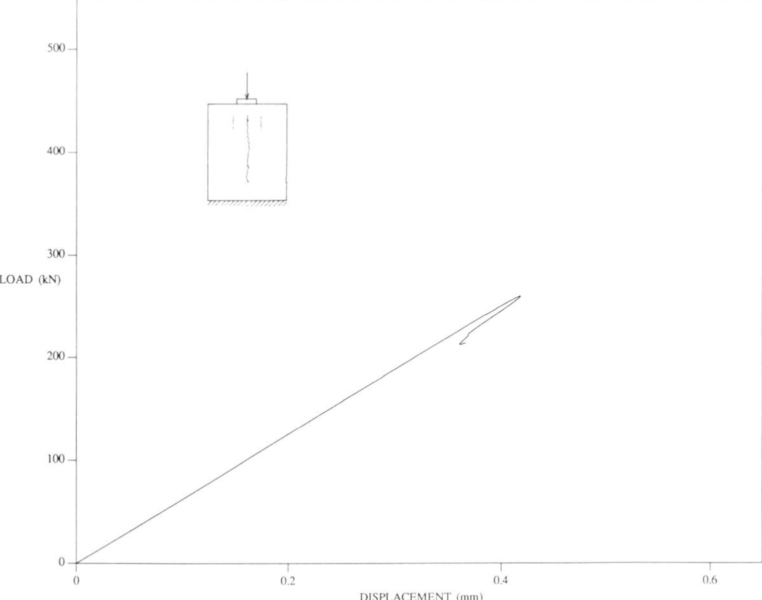

Fig. 9. Discrete crack analysis of masonry wall tested by Ali and Page [26]. Relation between vertical load and vertical displacement.

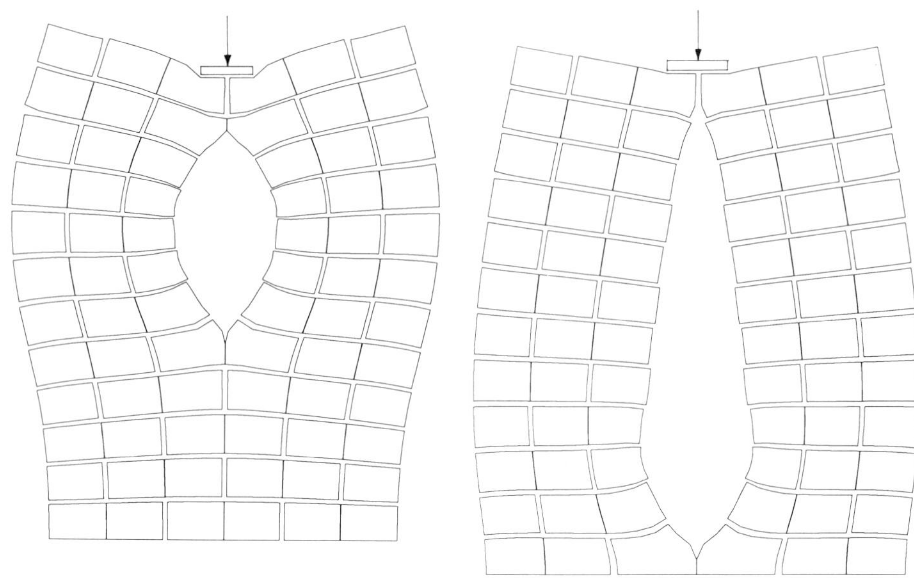

Fig. 10. Discrete crack analysis of masonry wall tested by Ali and Page [26]. Incremental displacement fields, (a) near the peak load in Fig. 9; (b) beyond the peak load in Fig. 9.

10. Conclusions

- The gradual transition from fixed, fixed multi-directional to rotating smeared crack approaches becomes clear by using a strain-decomposition approach in combination with proper variations of shear retention functions, softening laws and crack inclination angles.
- When smeared approaches are used, the coaxial rotating crack concept and the fixed crack concept with negligible shear retention provide the best possible results. Fixed cracks with significant shear retention lead to severe stress rebuild after cracking and overstiff response.
- A general deficiency of the smeared approach, either fixed, multi-directional or rotating, is the danger of stress-locking. This is due to the fact that geometrical *dis*-continuities are modeled using the assumption of displacement continuity. The discrete crack concept involves a discontinuum and does not suffer from stress-locking.
- The use of interface elements as potential discrete cracks is versatile for cases where the fracture path can be pre-assumed, and cases where certain fracture paths prevail (e.g. joints in masonry).

Acknowledgements

The computations have been carried out with the DIANA finite element code of the TNO Institute for Building Materials and Structures. The research has been assisted financially by the Centre for Civil Engineering Research, Codes and Specifications (CUR) and by the Netherlands Technology Foundation (STW). Major parts of the results have been obtaind in a

STW research project under supervision of Professor J. Blaauwendraad of Delft University of Technology.

References

1. S. Balakrishnan and D.W. Murra, IABSE Reports 54, Colloquium on Computation Mechanics of Reinforced Concrete, Delft University (1978) 393–404.
2. K. William, E. Pramono and S. Sture, in *Fracture of Concrete and Rock*, S.P. Shah and S.E. Swartz (eds.), SEM, Bethel (1987) 192–207.
3. F. Barzegar, *ASCE Journal of Structural Engineering* 115 (3) (1989) 647–665.
4. M.A. Crisfield and J. Wills, *ASCR Journal of Engineering Mechanics* 115 (3) (1989) 578–597.
5. D. Ngo and A.C. Scordelis, *Journal of the American Concrete Institute* 64 (14) (1967) 152–163.
6. R.E. Goodman, R.L. Taylor and T.L. Brekke, *ASCE Journal of Solid Mechanics and Foundation Division* 94 (3) (1968) 637–659.
7. J.G. Rots, Computational Modeling of Concrete Fracture, Dissertation, Delft University of Technology, Faculty of Civil Engineering (1988).
8. J.-M. Hohberg and H. Bachmann, in *Numerical Methods in Geomechanics*, G. Swoboda (ed.), Balkema Rotterdam (1988) 829–834.
9. R. de Borst and P. Nauta, *Engineering Computations* 2 (1985) 35–46.
10. J.G. Rots, P. Bauta, G.M.A. Kusters and J. Blaauwendraad, *HERON* 30 (1) (1985) 1–48.
11. Z.P. Bazant, *ASCE Journal of Engineering Mechanics* 109 (3) (1983) 849–865.
12. A.K. Gupta and H. Akbar, *ASCE Journal of Structural Engineering* 110 (8) (1984) 1735–1746.
13. A. Hillerbordg, M. Modeer and P.E. Petersson, *Cement and Concrete Research* 6 (6) (1976) 773–782.
14. Z.P. Bazant and B.H. Oh, *RILEM Materials and Structures* 16 (93) (1983) 155–177.
15. H.W. Reinhardt, H.A.W. Cornelissen and D.A. Hordijk, *ASCE Journal of Structural Engineering* 112 (11) (1986) 2462–2477.
16. J.G. Rots and R. de Borst, *International Journal of Solids and Structures*, accepted for publication.
17. M. Arrea and A.R. Ingraffea, Report 81–13, Department of Structural Engineering, Cornell University, Ithaca, New York (1982).
18. A.S. Kobayashi, M.N. Hawkins, D.B. Barker and B.M. Liaw, in *Application of Fracture Mechanics to Cementitious Composites*, S.P. Shah (ed.), Nijhoff Publishers, Dordrecht (1985) 25–50.
19. Y.S. Jenq and S.P. Shah, *International Journal of Fracture* 38 (1988) 122–142.
20. J.G.M. van Mier and M.B. Nooru-Mohamed, in *Fracture Toughness and Fracture Energy-Test Methods for Concrete and Rock*, H. Mihashi et al. (eds.), Tohoku University, Sendai, Japan (1988) 433–447.
21. M. Suidan and W.C. Schnobrich, *ASCE Journal of the Structural Division* 99 (10) (1973) 2109–2122.
22. J.G. Rots and R. de Borst, *ASCE Journal of Engineering Mechanics* 113 (11) (1987) 1739–1758.
23. R. de Borst, *Computers and Structures* 25 (2) (1987) 211–224.
24. J. Lubliner, J. Oliver, S. Oller and E. Onate, *International Journal of Solids and Structures* 25 (3) (1989) 229–326.
25. J.G. Rots, in *Fracture Toughness and Fracture Energy-Test Methods for Concrete and Rock*, Tohoku University, Sendai, Japan (1988) 285–300.
26. Sk.S. Ali and A.W. Page, *ASCER Journal of Structural Engineering* 114 (8) (1988) 1761–1784.

International Journal of Fracture **51**: 61–72, 1991.
Z.P. Bažant (ed.), Current Trends in Concrete Fracture Research.
© 1991 *Kluwer Academic Publishers. Printed in the Netherlands.*

Code-type formulation of fracture mechanics concepts for concrete

H.K. HILSDORF[1] and W. BRAMESHUBER[2]
[1]*Institut für Massivbau und Baustofftechnologie, University of Karlsruhe, Germany;*
[2]*BUNG consultants, Heidelberg, Germany*

Received 11 June 1990; accepted 11 November 1990

Abstract. The new model code for the design of concrete structures of the Comite Euro-International du Béton (CEB) includes extensive information on constitutive relations for concrete and reinforcing steel. In this model code relations are also proposed to predict fracture properties of concrete on the basis of fracture mechanics concepts. In particular fracture energy G_F is given as a function of concrete grade, maximum aggregate size and temperature. In addition, bilinear stress-strain and crack opening relations are presented. In this paper these relations are verified on the basis of theoretical considerations and available experimental data.

1. Introduction

In 1978 the Comité Euro-International du Béton published the CEB–FIP Model Code 1978 for the design of reinforced concrete structures [1]. This model code served as a basis for various national codes and in particular for the Eurocode EC2 'Design of Concrete Structures' [2]. The CEB–FIP Model Code 1978 gave only limited information on material properties and on constitutive relations for concrete with the exception of creep and shrinkage which has been dealt with in an appendix.

In the process of revising the CEB–FIP Model Code it became apparent that constitutive relations both for concrete and for reinforcing steel should be an integral part of a modern code for reinforced and prestressed concrete structures. Such relations are urgently needed in particular for non-linear analyses and finite element calculations. Therefore, in the predraft of the CEB–FIP Model Code 1990 (MC 90) [3] a Section 2.1 'Concrete – Classification and Constitutive Relations' has been included which gives information on the following concrete properties:

— compressive strength, tensile strength and fracture energy;
— strength under multiaxial states of stress;
— stress-strain relations and stress-crack opening relations;
— effects of stress and strain rate on strength and deformation properties;
— effects of time on strength and deformation properties;
— effects of temperature on strength and deformation properties;
— transport of liquids and gases in hardened concrete.

The information given on fracture properties is based on fracture mechanics concepts, making use of the tremendous progress in this field during the past decade throughout the world and in various national organizations such as RILEM and ACI. Chapter 2.1 of the Predraft to MC 90 has been prepared primarily by the authors of this paper and by Dr. H.S. Müller, Bundesanstalt für Materialforschung und -prüfung, Berlin, under the auspices of CEB-Commission VIII 'Concrete Technology'. In addition, extended use has been made of the work of other groups within CEB.

In the following, relations for an estimate of fracture parameters as given in [3] as well as their justification are summarized. Since the publication of the predraft of MC 90, some minor changes in the constitutive relations have been made. They are taken into account in this paper.

2. Relations for fracture mechanics parameters given in MC 90

2.1. Parameters and input data

From the work of several investigators it follows that the behavior of concrete and reinforced concrete elements subjected to tensile stresses can be analyzed in a realistic way on the basis of the following characteristics of concrete [4], [5], [6]:

— the axial tensile strength f_{ct};
— the fracture energy G_F, defined as the energy required to propagate a tensile crack of unit area;
— stress-strain relations for increasing stresses up to the level of tensile strength and a limiting tensile strain;
— stress-crack opening relations.

Since the CEB-Model Codes are directed towards the designer, all constitutive relations have been formulated such that only parameters are used which are generally known to the designer at the stage of design. Therefore, for the prediction of fracture properties the following parameters have been taken into account:

— strength grade of the concrete expressed by its characteristic compressive strength f_{ck} [MPa];
— the maximum size of aggregates, d [mm];
— temperature of the ambient air in the range of $0°C < T < 80°C$.

The experimental data available were not sufficient to also take into account strain or stress rate effects, concrete age or the influence of sustained loads.

2.2. Relations for fracture energy

In the absence of experimental data for a particular concrete G_F may be estimated from

$$G_F = a_d \cdot f_{cm}^{0,7},\tag{1}$$

where G_F = fracture energy [Nm/m²]; $f_{cm} = f_{ck} + 8$ = mean compressive strength of concrete [MPa]; f_{ck} = characteristic compressive strength of concrete [MPa] defined as the 5 percent defective; a_d = coefficient to be taken from Table 1. It depends on the maximum aggregate size, d.

Table 1. Coefficient, a_d, to take into account the effect
of maximum aggregate size, d, on fracture energy G_F

d [mm]	a_d
8	4
16	6
32	10

2.3. Stress-strain and stress-crack opening relations for uniaxial tension

The following relations are given for the modulus of elasticity of concrete, E_c and for the mean
tensile strength f_{ctm}:

$$E_c = 10^4 \cdot f_{cm}^{1/3} \tag{2}$$

and

$$f_{ctm} = 0.30 \cdot f_{ck}^{2/3}. \tag{3}$$

For uncracked concrete a bilinear stress-strain relation as expressed by (4) and (5) and shown
in Fig. 1 may be used:

$$\sigma_{ct} = E_c \cdot \varepsilon_{ct} \quad \text{for } \sigma < 0.9 f_{ctm}, \tag{4}$$

$$\sigma_{ct} = f_{ctm} - \frac{0.1 f_{ctm}}{0.00015 - \dfrac{0.9 f_{ctm}}{E_c}} \cdot (0.00015 - \varepsilon_{ct}) \tag{5}$$

where σ_{ct} = tensile stress [N/mm^2]; ε_{ct} = tensile strain; E_c = tangent modulus of elasticity of
concrete [N/mm^2] to be estimated from (4); f_{ctm} = mean axial tensile strength of concrete
[MPa] to be estimated from (5).

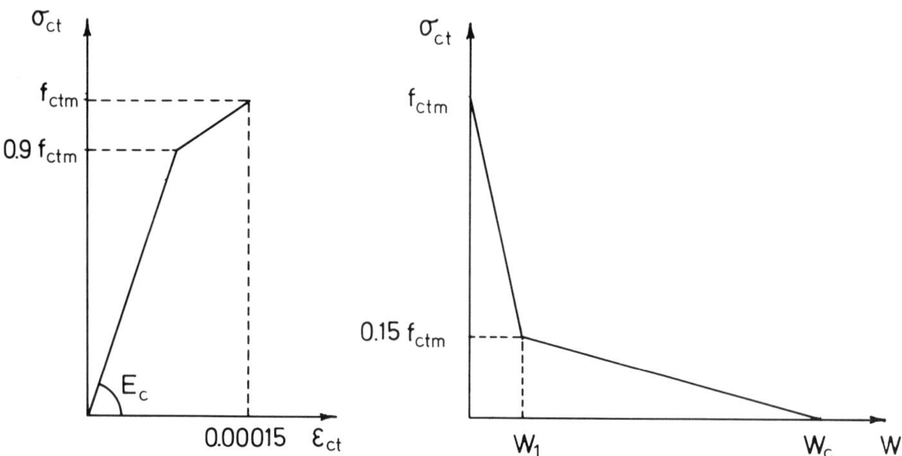

Fig. 1. Stress-strain and stress-crack opening diagram for uniaxial tension.

For cracked sections a bilinear stress-crack opening relation as described by (6), (7) and (8) and shown in Fig. 1 is proposed:

$$\sigma_{ct} = f_{ctm}\left(1 - 0.85\frac{w}{w_1}\right) \quad \text{for } 0.15f_{ctm} < \sigma_{ct} < f_{ctm}, \tag{6}$$

$$\sigma_{ct} = \frac{0.15f_{ctm}}{w_c - w_1}(w_c - w) \quad \text{for } 0 < \sigma_{ct} < 0.15f_{ctm} \tag{7}$$

and

$$w_1 = \frac{G_F - 22w_c(G_F/a_d)^{0.95}}{150(G_F/a_d)^{0.95}}, \tag{8}$$

where w_1 = crack opening at the nick [mm] as defined in Fig. 1; w_c = crack opening [mm] for $\sigma_{ct} = 0$; G_F = fracture energy acc. to (1); a_d = coefficient to be taken from Table 1.

The crack opening w_c at $\sigma_{ct} = 0$ depends on the maximum aggregate size and may be taken from Table 2.

Table 2. Crack opening w_c for $\sigma_{ct} = 0$

d_{max} [mm]	w_c [mm]
8	0.12
16	0.15
32	0.25

2.4. Effect of temperature

For the temperature range $0°C < T < 80°C$ the effect of temperature on fracture energy G_F may be estimated from (9) and (10)

$$\text{for dry concrete: } G_F(T) = G_F(1.07 - 0.0030T), \tag{9}$$

$$\text{for mass concrete: } G_F(T) = G_F(1.14 - 0.006T), \tag{10}$$

where $G_F(T)$ = fracture energy at temperature T; G_F = fracture energy at $T = 20°C$ from (1); T = temperature in [°C].

For the temperature range considered the temperature dependence of the uniaxial tensile strength f_{ctm} does not have to be taken into account. The influence of temperature on the modulus of elasticity may be estimated from (11)

$$E_c(T) = E_c(1.06 - 0.003T), \tag{11}$$

where $E_c(T)$ = modulus of elasticity at temperature T; E_c = modulus of elasticity at $T = 20°C$ from (4).

3. Verification of relations

3.1. Fracture energy – technological parameters

In order to evaluate the major parameters influencing fracture energy G_F the experimental data reported in [7], [8], [9], [10] have been studied carefully. Particular attention has been given to the results of round robin tests which were reported in [9]. Since the data given in [9] were partially incomplete, additional information has been obtained by direct contacts with the various investigators. This additional information as well as all other data needed for the evaluation are given in detail in [11]. From these experimental data the following technological parameters were found to be of particular significance for the fracture energy:

— compressive strength and water/cement ratio of the concrete;
— maximum aggregate size;
— concrete age.

In addition, geometrical parameters, in particular the depth of the ligament above a crack or notch are of significance. These effects will be dealt with in Section 3.2. The parameters given above also have been verified in [8] and [10] as being the most significant in influencing fracture energy.

Since water-cement ratio, concrete age and compressive strength of the concrete are interrelated and since the available data base was not sufficient to clearly distinguish between the effects of these parameters, only compressive strength and maximum aggregate size have been chosen as parameters for the Code prediction of fracture energy. For the evaluation the results of 36 experiments described in detail in [11] have been used. Figure 2 gives the relation between the mean concrete compressive strength f_{cm} and fracture energy G_F on a double logarithmic scale for maximum aggregate sizes, d, of

$$1 < d < \ 8 \, \text{mm},$$
$$12 < d < 20 \, \text{mm},$$
$$d = 32 \, \text{mm}.$$

The relation between G_F and f_{cm} is best documented for $12 < d < 20$ mm, $d_{av} = 16$ mm. For this aggregate size (1) with $a_d = 6.0$ for $d = 16$ mm results in a correlation coefficient $k = 0.83$.

Figure 2 also shows the well known tendency that fracture energy increases with increasing maximum aggregate size. The same trend is true for other fracture characteristics such as K_{Ic}. However, the data base is too small to establish safe predictions for a maximum aggregate size $d < 12$ mm and $d > 20$ mm. Therefore, the values of a_d for $d = 4$ mm and $d = 32$ mm given in Table 1 should be taken with caution. This is particularly true for $d = 32$ mm where only the result of one series of experiments is available. Therefore, additional experiments are required to ensure the relations between G_F and f_{cm} for max. aggregate sizes other than $12 < d < 20$ mm, and in particular for $d = 32$ mm. Where more accurate predictions are required, G_F should be determined experimentally, e.g. [12].

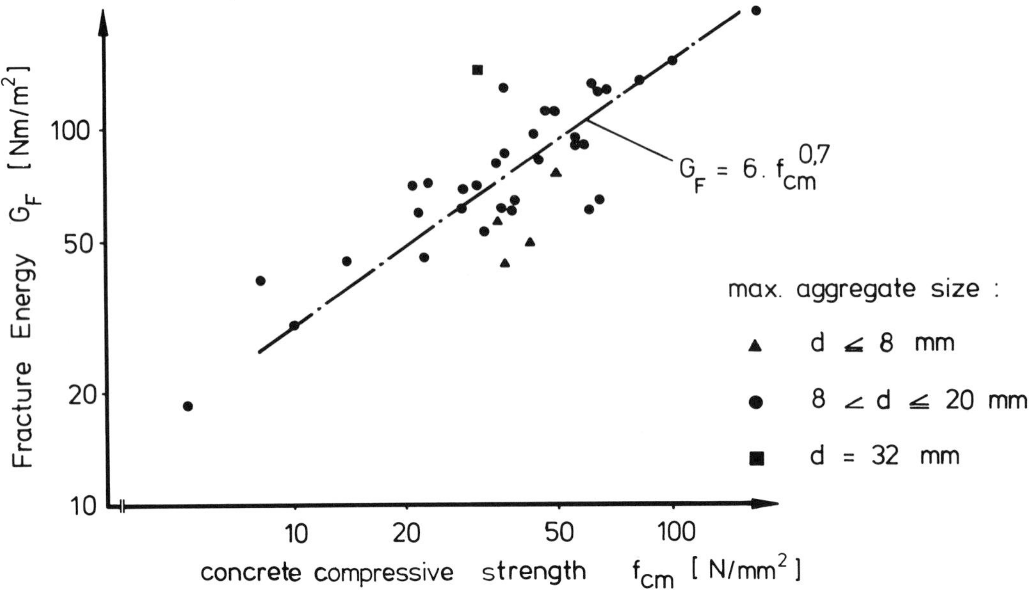

Fig. 2. Fracture energy and compressive strength of concrete.

In [7] and [15] it was shown that the characteristic length l_{ch} as defined in (12) is particularly suitable to describe the ductility and crack sensitivity of concrete:

$$l_{ch} = \frac{G_F \cdot E_{cm}}{f_{ctm}^2}, \tag{12}$$

where l_{ch} = characteristic length [mm]; E_{cm} = modulus of elasticity [N/mm²]; f_{ctm} = mean tensile strength of concrete [MPa].

From (1), (2), (3) and (12) a relation between l_{ch} and concrete compressive strength f_{cm} can be derived. It may be approximated by (13).

$$l_{ch} = 600a_d \cdot f_{cm}^{-0.3}. \tag{13}$$

Deviating from (3) it was assumed that $f_{ctm} = 0.30 f_{cm}^{2/3}$. Figure 3 shows the experimental values of l_{ch} as a function of f_{cm} for a maximum aggregate size $12 < d_{max} < 20$ mm. In contrast to fracture energy, the characteristic length decreases as the concrete compressive strength increases. Equation (13) describes the available experimental data reasonably well though the correlation coefficient $k = 0.72$ is lower than the corresponding value for G_F. This is not surprising since the prediction of l_{ch} from (13) also includes uncertainties in the estimate of tensile strength and modulus of elasticity.

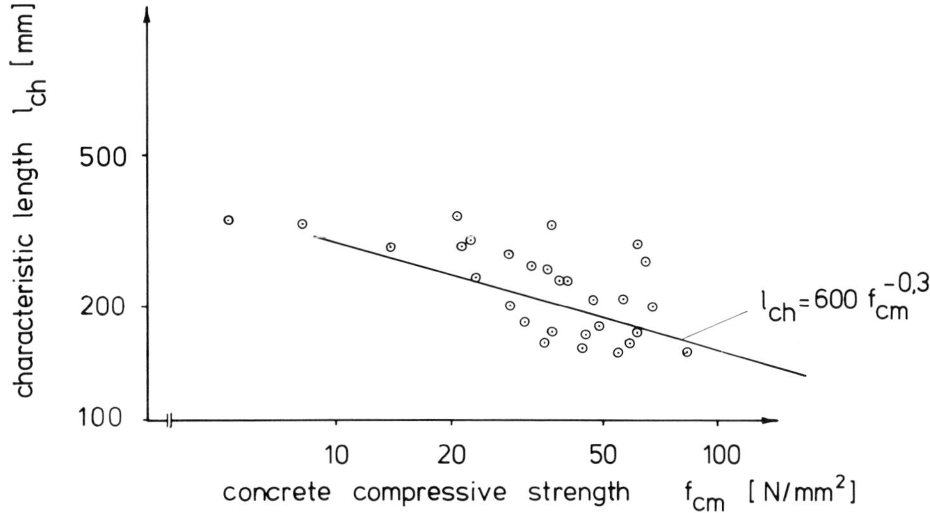

Fig. 3. Characteristic length and compressive strength of concrete.

3.2. *Fracture energy – size effects*

Various experiments [7–10] show that fracture energy G_F if determined experimentally according to [12] increases with increasing depth of the uncracked ligament. In [8] it was shown furthermore, that for a depth of the ligament larger than approximately 300 mm, fracture energy is little affected by a further increase of the ligament depth.

Various approaches have been proposed to take into account this size effect, in particular the size effect law developed by Bazant [e.g. 15]. Though the general validity of the size effect law is not questioned it appeared to be desirable to find a size independent approach to predict G_F for a Code type formulation irrespective of the inevitable errors which may be introduced by such a formulation. This is even more so since size effects on plain and reinforced concrete properties can be predicted even with a size independent G_F.

In [7, 16] the causes of the size dependence of G_F have been analyzed in more detail. In experiments on notched beams of different depth, however, with a constant ratio of notch depth/beam depth of 0.5 the crack propagation has been determined carefully, and the relative fracture energy G_F/G_{F0} required to propagate a crack has been determined as a function of crack length Δc. Figure 4 shows the result of this analysis. There, the relative fracture energy is given as a function of Δc for beams with uncracked ligament depths between 50 and 400 mm. Irrespective of the initial depth of the ligament the fracture energy G_F required to propagate the crack increased with increasing crack length Δc up to a crack depth of approximately 40 mm. For a further increase in crack length the fracture energy stayed constant at a level $G_F = G_{F0}$. Since the crack length at which a constant level of G_{F0} is reached is independent of the depth of the ligament the average value of G_F decreases as the depth of the ligament decreases. Table 3 summarizes the errors which occur if constant values of G_F valid for beams with a ligament depth >800 mm are applied to beams with a smaller ligament depth. These errors are generally less than 20 percent. Thus they are small compared to other experimental errors as shown in [7, 16]. They are in the range of size dependence of the compressive strength of concrete.

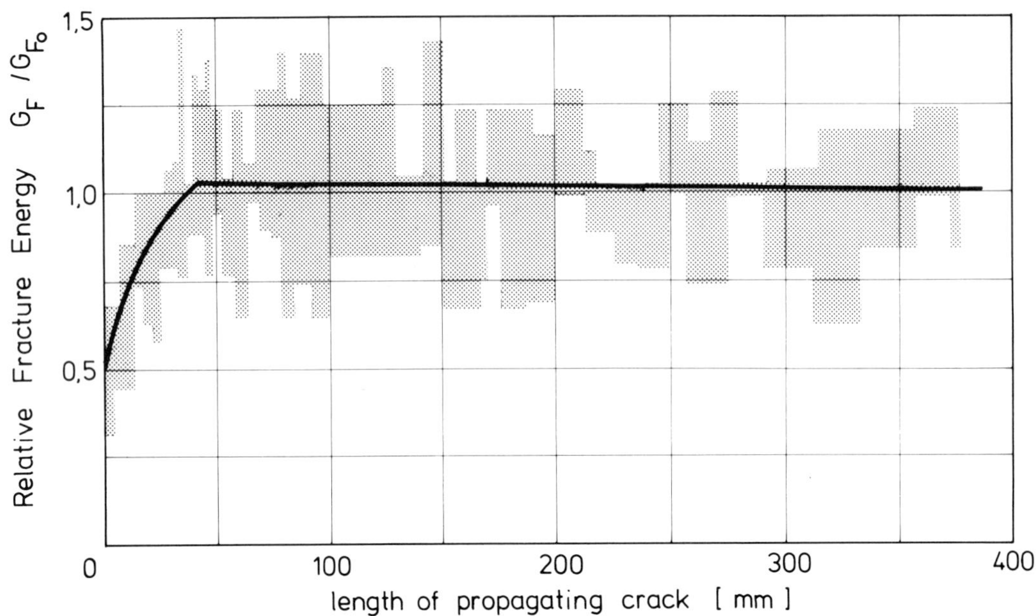

Fig. 4. Fracture energy as a function of length of propagated crack.

Table 3. Fracture energy G_F as a fraction of fracture energy of a deep beam G_{F0}; $a/d = 0.5$

depth of beam	G_F/G_{F0}	
(mm)	concrete $d_{max} = 32\,mm$	mortar $d_{max} = 4\,mm$
100	1.16	1.08
200	1.08	1.04
400	1.04	1.02
800	1.02	1.01
1000	1.00	1.00

3.3. Fracture energy – temperature effects

The data base to evaluate the effect of temperature in the range of $0°C < T < 80°C$ on fracture energy is small [7, 15]. Figure 5 summarizes the available results. There, fracture energy G_F at a given temperature is expressed as a fraction of G_F at $T = 23°C$ and plotted as a function of the temperature at testing, T. From Fig. 5 it follows that fracture energy decreases linearly with increasing temperature. In addition, the moisture state of the concrete is of significance: dry concretes are less temperature sensitive than wet concretes.

Figure 5 also shows the relations given in MC 90 to describe the effect of temperature on G_F (9–10). They are in close agreement with the available test data for a temperature range $0°C < T < 80°C$.

In [7, 16] the theoretical basis for a linear relationship between G_F and T in the above temperature range has been given. It is based on the relation between potential energy of bonding and temperature. As shown in [7, 16], (9) and (10) allow extrapolations up to

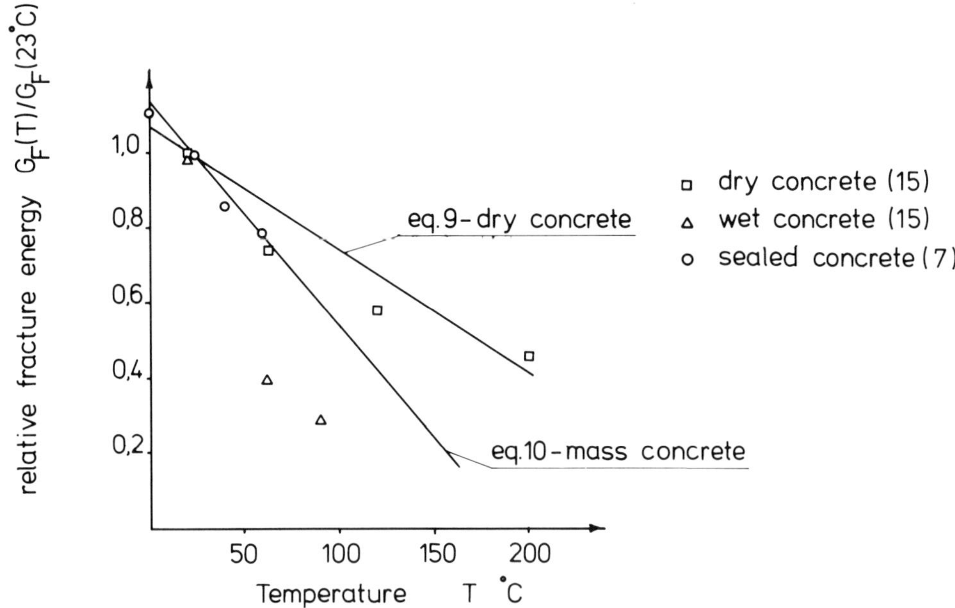

Fig. 5. Effect of temperature on fracture energy of concrete.

temperatures as low as $-170°C$. For $T > 80°C$ the actual relation between G_F and T is non-linear so that an extrapolation of (9) and (10) is no longer permissible. In [15] a model for the relation between G_F and T is given which is based upon activation energy considerations. Both models differ little in the temperature range $0°C < T < 80°C$. However, the model given in [15] does not allow extrapolations to lower temperatures since such an extrapolation would result in considerable overestimates of G_F at lower temperatures [7, 16].

3.4. Stress-strain and stress-crack opening relations

The description of the stress-strain properties of concrete subjected to tensile stresses is based primarily upon a proposal initially made by Petersson [13]. This proposal consists of a linear stress-strain relation of uncracked concrete and a linear stress-crack opening relation for cracked concrete. In reality the stress-strain relation for uncracked concrete is not entirely linear. Therefore, a bilinear function expressed by (4) and (5) has been chosen to take into account the non-linear behavior at stresses $\sigma_{ct} > 0.9 f_{ctm}$. A constant strain at maximum stress $\varepsilon_{ctmax} = 0.00015$ has been assumed since no systematic effects on this parameter could be found.

A variety of formulations have been tested to describe the strain softening behavior of the cracked concrete. Of particular significance is the question to which extent the stress-crack opening relations found experimentally could be simplified for a code-type formulation. Therefore, various calculations have been carried out which were based on the fictitious crack model described in [13].

From the literature and in particular from the results of the aforementioned round robin tests load-deflection relations for various types of concretes and specimen geometries have been taken. Together with other concrete properties, in particular modulus of elasticity, tensile strength and fracture energy, theoretical load-deflection relationships have been calculated based on bilinear stress-crack opening relations of the shape shown in Fig. 1. The parameter w_c, i.e. the maximum crack opening at zero stress, and the crack opening at the nick w_1 of the bilinear $\sigma - w$ relation have been varied in order to obtain an acceptable agreement between theoretical and experimental load-deflection curves. The following simplifying assumptions were made as a result of these trial calculations:

1. The maximum crack opening at zero stress has little effect on the calculated load-deflection relations. It, therefore, would be an unnecessary complication to express w_c as a direct function of G_F or f_{ctm}. However, the maximum crack opening increases with increasing maximum aggregate size as expressed by Table 2.
2. The agreement between experimental and theoretical load-deflection relations is strongly influenced by the slope of the initial part of the bilinear $\sigma - w$ relation. However, variations of the nick in the $\sigma - w$ relation are less significant. Therefore, a value of $\sigma_{ct}(w_1) = 0.15 f_{ctm}$ may be employed.

From the condition that

$$G_F = \int_{w=0}^{w=w_c} \sigma_{ct}(w) \cdot dw$$

the value of w_1 can be calculated:

$$w_1 = \frac{G_F - \dfrac{w_c}{2} \cdot \sigma_{ct}(w_1)}{0.5 f_{ctm}} \tag{15a}$$

and

$$w_1 = \frac{G_F - \dfrac{w_c}{2} \cdot 0.15 f_{ctm}}{0.5 f_{ctm}}. \tag{15b}$$

Expressing f_{ctm} in terms of fracture energy on the basis of (1) and the modified (3), (8) is obtained which gives the crack opening at the nick in terms of fracture energy G_F and max. aggregate size. Thus the stress-crack opening behavior of concrete can be described by (6)–(8).

These equations depict the well-known characteristics for concrete loaded in tension, in particular

— decreasing non-linearity of the stress-strain relations with increasing compressive strength;
— decreasing slopes of the stress-crack opening relations with increasing compressive strength.

In Fig. 6 load deflection curves determined experimentally and reported in [7] are compared to theoretical load-deflection curves calculated on the basis of the fictitious crack model [14] and (1)–(8). Acceptable agreement has been obtained.

In addition to the bilinear functions for the stress-crack opening relations given above also continuous functions have been developed at our institute. Similar functions have been reported

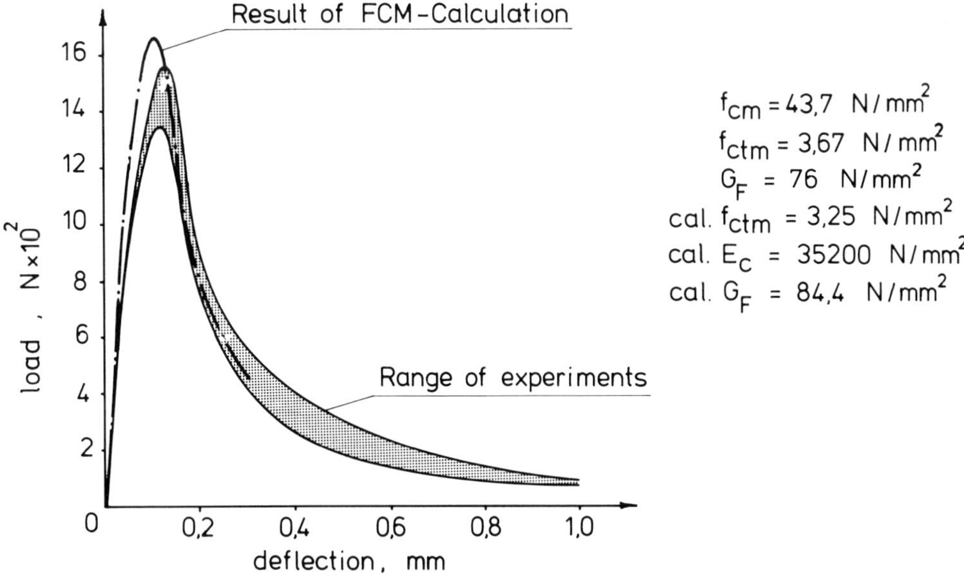

Fig. 6. Comparison of experimental and calculated load-deflection curves of plain concrete beams, 3-point-bending. $1 \times d \times b = 500 \times 100 \times 100$ mm; notch depth: 50 mm; max. aggregate size: 16 mm.

in the literature [17, 18]. It also has been proposed to express w, in terms of G_F/f_{ctm} [19]. However, there is not sufficient experimental evidence based on tests on concentrically loaded concrete specimens that such interrelations exist and are of major significance. It must be pointed out in this context that the formulations given above are primarily based on load-deflection measurements of notched beams. Though comparison with the few experimental data from concentric tension tests indicates acceptable agreement, further experimental data in this field are urgently needed.

4. Summary and conclusions

In the CEB–FIP Model Code 1990 various constitutive relations to describe the properties of concrete are given. This section includes information on the fracture properties and stress-deformation characteristics of concrete loaded in tension. Among the fracture parameters under discussion fracture energy G_F, bilinear stress-strain relationships for the uncracked concrete and bilinear stress-crack opening relations for the cracked concrete have been chosen. Despite the fact that only parameters generally known to the designer, i.e. strength grade and maximum aggregate size, have been chosen as input data, acceptable agreement between prediction and experimental results has been obtained.

It is considered a major breakthrough that fracture mechanics data are included in an international concrete code, and it is hoped that this new approach will open new avenues for more realistic ways in non-linear structural analysis. But in deriving these code-type formulations of concrete fracture properties the lack of experimental studies oriented towards practical engineering applications became evident. It is hoped that in the planning of future research this gap in our knowledge will be kept in mind.

References

1. CEB–FIP Model Code for Concrete Structures, Comité Euro-International du Béton (CEB) Lausanne, 1978.
2. Eurocode No. 2, Design of Concrete Structures, Part 1, General Rules and Rules for Buildings, Final Draft, December 1988.
3. CEB–FIP Model Code 1990, First Predraft 1988, Bulletin d'Information No. 190a, 190b, Comité Euro-International du Béton (CEB), Lausanne.
4. A. Hillerborg, in *Fracture Mechanics of Concrete*, G.C. Sih, (ed.), Martinus Nijhoff Publishers (1983).
5. Z.P. Bazant and R. Gettu, in *Fracture of Concrete and Rock: Recent Developments*, S.P. Shah, S.E. Swartz, B. Barr (eds.), Elsevier Applied Science (1989).
6. M. Puche, *Rißbreitenbeschränkung und Mindestbewehrung bei Eigenspannungen und Zwang*, Deutscher Ausschuß für Stahlbeton, Heft 396, Beuth Verlag, Berlin (1988).
7. W. Brameshuber, Bruchmechanische Eigenschaften von jungem Beton, Schriftenreihe des Instituts für Massivbau und Baustofftechnologie, Universität Karlsruhe, Heft 5 (1988).
8. F.H. Wittmann, P.E. Roelfstra, H. Mehashi, Yiun-Yuang Huang, Xin-Hua Zhang and N. Nomura, *Materials and Structures* 20 (1987), *Proceedings of the 9th Annual Meeting of Japan Concrete Institute*, Vol. 9–2 (1987).
9. A. Hillerborg, *Materials and Structures* 18, No. 107 (1985).
10. P. Nallathambi, B.L. Karihaloo and B.S. Heaton, *Cement and Concrete Research* 15, No. 1 (1985).
11. W. Tschupak, Der Einfluß der Probengröße auf die Bruchenergie von Mörtel und Beton, Diplomarbeit am Institut für Massivbau und Baustofftechnologie, Universität Karlsruhe (1985).
12. RILEM Draft Recommendation: TC-50 FMC Fracture Mechanics of Concrete, *Material and Structures* 18, No. 106, Paris (1985).
13. P.-E. Petersson, Crack growth and development of fracture zones in plain concrete and similar materials, Division of Building Materials, Lund Institute of Technology, Report TVBM-1006, Sweden (1981).
14. W. Brameshuber and H.K. Hilsdorf, in *Proceedings of the International Conference of Fracture and Damage of Concrete and Rock*, Vienna (1988).
15. Z.P. Bazant, in *SEM/RILEM International Conference on Fracture of Concrete and Rock*, S.P. Shah and S.E. Swartz (eds.), Houston, Texas (1987).
16. W. Brameshuber and H.K. Hilsdorf, in *International Conference on Fracture of Concrete and Rock*, S.P. Shah and S.E. Swartz (eds.), Houston, Texas (1987).
17. *International Conference on Recent Developments of Fracture Mechanics of Concrete and Rock*, S.P. Shah, S.E. Swartz and B.I.G. Barr (eds.), Elsevier Publishers, Cardiff (1989).
18. A.M. Alvaredo and R.J. Torrent, *Materials and Structures* 20, No. 120, Paris (1987).
19. A. Hillerborg, Private communication (1988).

Author's Note Added in Proof

In the revised version of MC 90 to be published in 1991 the expressions (1) to (11) as well as Tables 1 and 2 have been slightly altered in order to make these relations dimensionally compatible.

International Journal of Fracture **51**: 73–92, 1991.
Z.P. Bažant (ed.), Current Trends in Concrete Fracture Research.
© 1991 *Kluwer Academic Publishers. Printed in the Netherlands.*

The collapse of the Schoharie Creek Bridge:
a case study in concrete fracture mechanics

DANIEL V. SWENSON[1] and ANTHONY R. INGRAFFEA[2]
[1]*Department of Mechanical Engineering, Kansas State University, Manhattan, Kansas 66506, USA;*
[2]*School of Civil Engineering and Program of Computer Graphics, Cornell University, Ithaca, New York 14853, USA*

Received 1 June 1990; accepted 1 November 1990

Abstract. On April 5, 1987, the New York State Thruway bridge over Schoharie Creek collapsed without warning. The primary cause of failure was scour beneath a plain concrete pier footing. However, a necessary secondary cause was unstable propagation of a single crack in the pier. Conditions for initiation of the curvilinear crack are first evaluated. It is concluded that about 28 feet of scour had to occur to initiate stable process zone formation at the point of initiation, but that at least 44 feet was required to cause unstable cracking. Simulation of propagation was studied using discrete representation in a finite element model and nonlinear fracture mechanics. About 5 feet of propagation was necessary to transition from nonlinear to LEFM. Good agreement was found between observed and predicted final crack trajectories, and load redistribution in the bridge structure was determined to have been a necessary part of the failure process. Discussions concerning the application of the finite element method to crack initiation problems and the use of the size effect to estimate failure conditions in large, plain concrete structures are also presented.

1. Introduction

On April 5, 1987, the New York State Thruway bridge over Schoharie Creek collapsed without warning causing the death of ten people. Figure 1 shows an elevation of the multiple span structure and Fig. 2 is an aerial view of the bridge a few hours after collapse. The New York State Thruway Authority commissioned a failure investigation which resulted in three excellent reports [1, 2, 3] which detailed the history of the bridge, site geology and hydrology, post-failure site investigations, materials testing, foundation and structural analyses.

In particular, the final report [3] contained a discussion describing the sequence of collapse and determined a specific cause of failure. It also contained a detailed description of the structural analyses performed to support the determination of events and the cause of failure. Significantly, this report is one of the first to use results [4, 5] of research into the fracture mechanics of concrete to support their finite element analyses. An essential component of the investigation included finite element analyses of one of the bridge piers. These analyses were

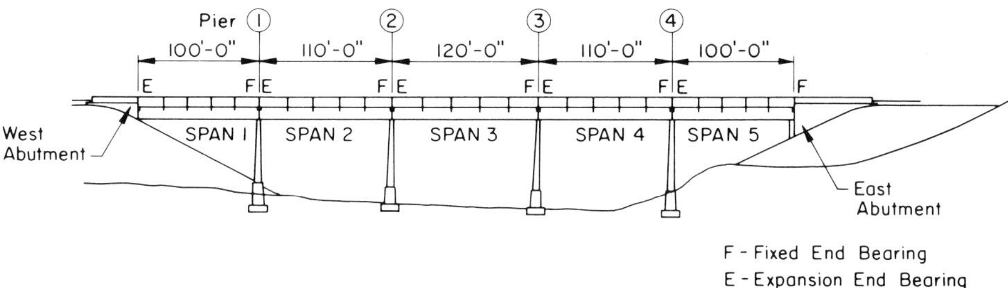

Fig. 1. Elevation of Schoharie Creek Bridge. From [3].

Fig. 2. Aerial view of bridge after failure. From [3].

guided by suspected scouring under the pier, which led to unstable cracking of a minimally reinforced section of the pier and precipitated collapse of the bridge. The analyses focused on elastic stresses and material properties necessary to initiate cracking, but did not attempt to determine stability of the crack nor attempt to simulate crack propagation.

The purpose of this paper is to investigate further, using nonlinear and linear fracture mechanics, conditions for initiation, stability, and trajectory of the crack which precipitated failure. This question is important not only to understand completely why the Schoharie Creek bridge collapsed, but also to propose preventive measures to safeguard many bridges with similar designs. Our motivation is in part an attempt to apply to a problem of practical significance the results of recent research into fracture mechanics of concrete.

The next section contains a detailed review of the falure analysis contained in [3], with emphasis on the finite element methodology and fracture mechanics in concrete. The following section describes the different technical approach taken in the present study and the fracture mechanics models used in our simulations. Following this, we present our results and discuss and compare them with those presented in [3]. Finally, in our conclusions, we note that although the finite element analyses in [3] were adequate to support identifying the cause of failure, they did not adequately address issues of stress concentration and crack stability, probably substantially underestimated the length of scour required to trigger the final fracture, and misconstrued some research results in the fracture mechanics of concrete.

2. Review of failure investigation

In this section we will summarize the findings presented in [3] regarding the sequence of collapse, the cause of failure, and the supporting finite element analyses. The collapse began with the apparently sudden rupture of the plinth and corresponding downward movement of the south end of pier three. Figure 3 shows the initial configuration of pier three, while Fig. 4 shows east and west views of the plinth of the failed pier taken during the site investigation. Since all of the spans of the bridge were simply supported, loss of support for the south girders of spans three and four resulted in collapse of the bridge. Further observations supporting the conjectured sequence of collapse are definitively documented in [3].

The apparently sudden rupture of the plinth of pier three '...occurred as a result of extensive undermining of the footing under pier three due to scour' [3]. The settlement due to undermining is clearly evident in Fig. 4. Undermining of the pier caused redistribution of the stresses in the pier, and in particular, apparently generated sufficient tensile stress in the top of the plinth near the base of the north column (point A in Fig. 3) to initiate the crack shown in Fig. 4. Two points must be noted: First, the plinth was not structurally reinforced; original design drawings show it to contain only shrinkage and temperature reinforcement. Second, in 1955, about a year after opening of the bridge, nearly vertical cracks were observed in the plinths of all the piers. In pier three, cracking was observed to begin at point B in Fig. 3. These cracks were attributed to excessive tensile flexural stresses due to support pressures on the base of the footing. Plinth reinforcement in Fig. 3 was added in 1957. The steel in the plinth reinforcement was not continuous through the base of the columns; it was connected to the plinth through a set of vertical dowels the first of which was located 20 inches from the inside column face. The dowels were intended to act as shear studs transmitting bending stresses from the plinth to the reinforcement.

The failure analysis in [3] focused on conditions necessary to initiate cracking. A series of two- and three-dimensional analyses was performed on pier three. These included calculations before and after addition of the plinth reinforcement, but with continuous soil support, and analyses with a variable amount of scour. The key result of these calculations is shown in Fig. 5. The maximum calculated stress at the top of the unreinforced plinth at the location of the

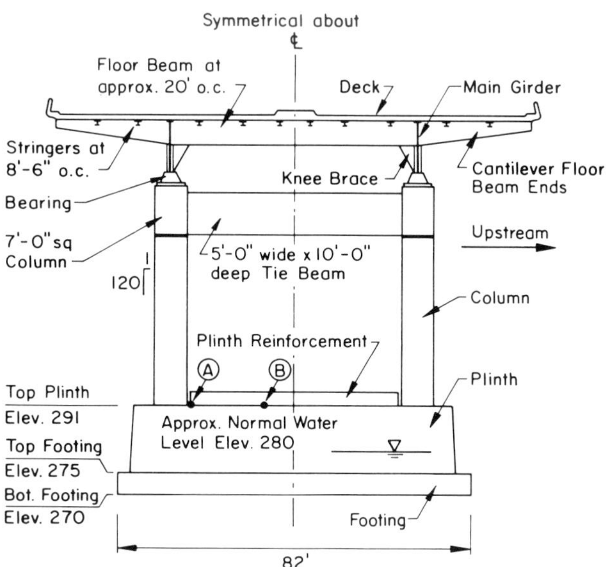

Fig. 3. Initial configuration of Pier Three, looking East. After [3].

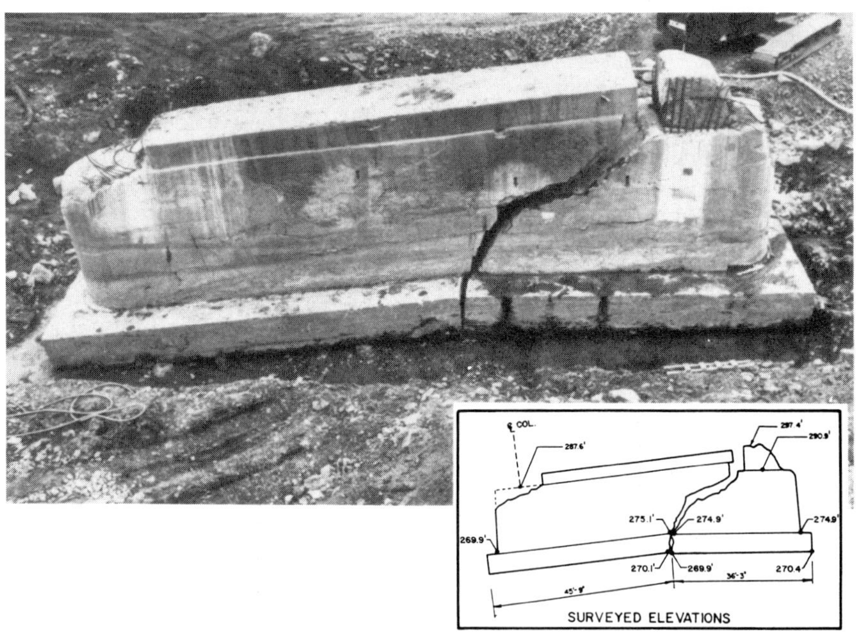

Fig. 4. (a) Photo of plinth of Pier Three after failure, looking East. (b) Photo of plinth of Pier Three after failure, looking West. Both from [3].

first set of dowels was computed to be 210 psi, when 28 feet of scour was assumed and soil pressures at the downstream end of the footing were allowed to be as high as 2 ksf in tension. Parameter studies showed that this maximum stress depended on the length of scour as well as the allowable tensile stresses in the soil. For 28 feet of scour, the variation in maximum

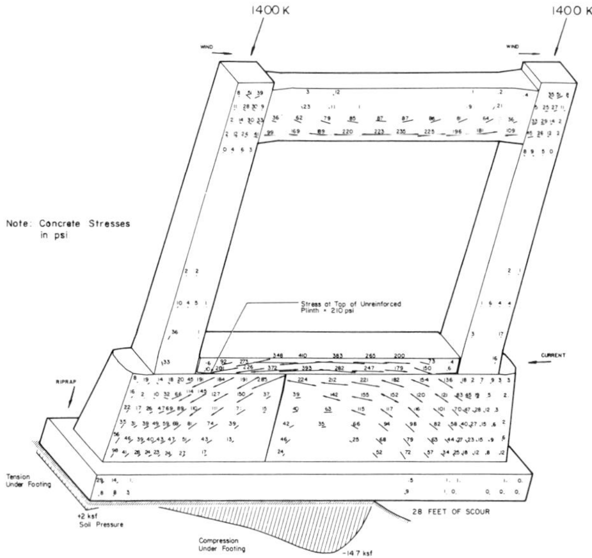

Fig. 5. Predicted stresses in Pier Three and soil after 28 feet of scour. From [3].

stress with tensile strength in the soil varying from 0 to 2 ksf was from 180 to 215 psi, respectively.

Split cylinder tests were performed on cores taken directly from pier three after failure, and these yielded an average splitting tensile strength of 620 psi. However, arguments were presented, based on research into size effect in the fracture of concrete, that the effective modulus of rupture of the concrete in pier three was only about 210 psi.

Two fundamental problems with this study are:

1. The finite element results are inherently mesh dependent. The reason for this is as follows. The location at which the maximum stress is computed in the mesh is modeled as a reentrant corner. As such, all stress components at this point are theoretically infinite. Had a mesh convergence study been performed (apparently the analysts used only one mesh configuration for their 2-D and 3-D analyses), the results would have been increasingly high values of maximum stresss with decreasing element size in the region of maximum stress. Therefore any stress could have been computed at this point, as quantitatively illustrated in the next section.

2. Significantly, the concept of size-effect in fracture mechanics was employed in [3]. It was reasoned correctly that the splitting tensile strength results, obtained from small diameter cores, substantially overestimated the actual modulus of rupture, f_r, which would have been measured on a beam with the depth of the plinth. Bazant's size effect model [4] was referenced and used approximately to estimate a proper modulus of rupture. However, the misapplication of this powerful and useful method occurred when the modulus of rupture and local tensile strength were used interchangeably. Since their finite element model was predicting a local tensile stress of only about 210 psi, the analysts in [3] were seeking a reduced local tensile strength to initiate cracking. The problem here is that size effect does not reduce the tensile strength at point A. This is explained clearly in Fig. 6(a) which is taken directly from [6], a paper in this special issue. It shows that in a modulus of rupture test, one still has to obtain the direct tensile strength, f_t, at the point where the cracking initiates. A hypothetical application of size-effect to

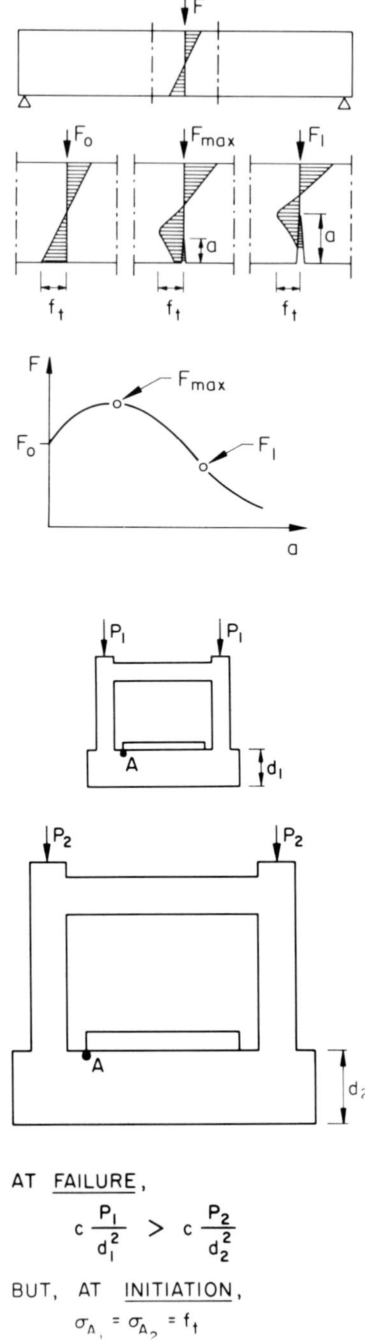

AT FAILURE,
$$c\,\frac{P_1}{d_1^2} \;>\; c\,\frac{P_2}{d_2^2}$$

BUT, AT INITIATION,
$$\sigma_{A_1} = \sigma_{A_2} = f_t$$

Fig. 6. Stress distributions and corresponding load history during stages of a modulus of rupture test. After [6].

the Schoharie Creek Bridge is presented in Fig. 6(b). It shows two sizes of pier 3. The relationship predicted by size-effect between normalized *failure* loads is indicated; however, the crack initiation stress at point A, f_t, is the same for both sizes.

This point is further emphasized by the form of Bazant's model:

$$f_r = \frac{f_t}{\sqrt{1 + k(d/d_a)}},\tag{1}$$

where d = specimen depth; d_2 = aggregate size; k = an empirical constant; and where f_t, the stress at crack is initiation, is also a constant. This is the reason we used 550 psi, an estimate of f_t from the measured splitting tensile strength (the average measured compressive strength was 6,200 psi), in our analyses. It is this value of stress that would have *initiated* the cracking process in the plinth, although attainment of this stress is not necessarily synonymous with unstable crack propagation and failure.

3. Technical approach

3.1. Nonlinear fracture mechanics

A basic assumption of LEFM is that the process zone is small compared to all other characteristic lengths in the structure. Research [7, 8, 9] has shown that in concrete the process zone is often relatively large (on the order of inches). The behavior occurring in the process zone is strain-softening of the concrete [9] and illustrated in Fig. 7. Ahead of the process zone, the concrete behaves elastically. In the process zone, the concrete is softening, and the greater the crack-opening-displacement (COD) the less stress is transferred across the process zone. Finally, there is a region in which there is no stress on the crack face. This behavior has important consequences. First, during initiation, the entire crack will be a process zone and clearly linear elastic fracture mechanics is not applicable [9]. Additionally, the process zone size is a function not only of the concrete material properties, but also of the structure shape and boundary conditions. For instance if a uniform specimen is uniformly loaded in tension, the process zone can exist across the entire specimen [7]. However, if the crack is growing in a region with a strong stress gradient (such as a beam in bending), the process zone for the same material can be relatively short [7, 9]. Therefore, the crack length at which LEFM becomes applicable in concrete cannot be stated *a priori*, but must be determined by analysis.

The model we use to represent this experimentally observed behavior is the generalized Dugdale-Barenblatt approach developed by Hillerborg [8] and used in a number of other

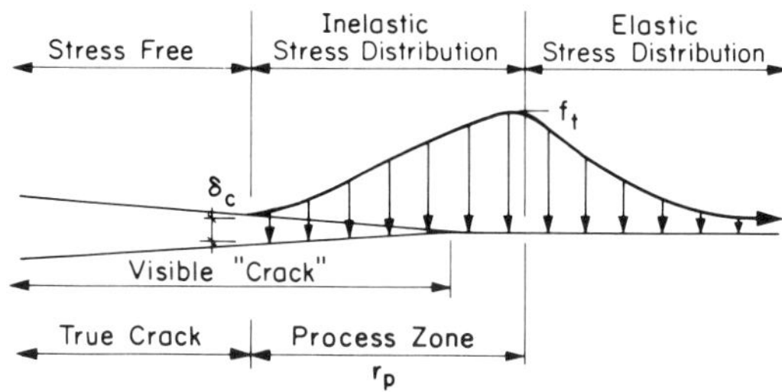

Fig. 7. Schematic of cohesive process zone concept in concrete. From [9].

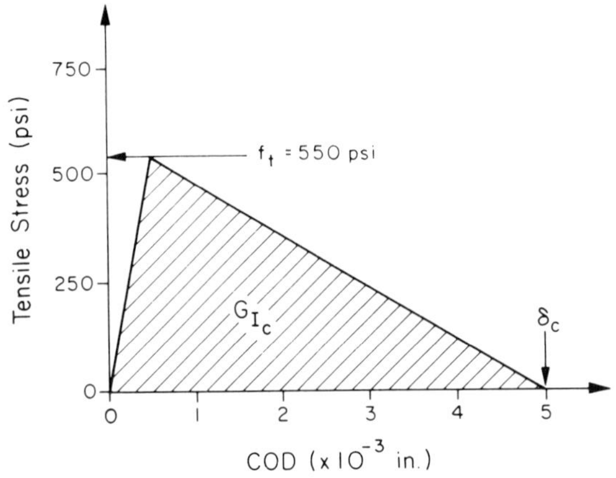

Fig. 8. Process zone constitutive model used in present analyses.

applications [10, 11, 12]. In this approach a discrete process zone is modeled by specifying a stress-vs-COD relationship of the type shown in Fig. 8. This relationship is characterized by a tensile strength, f_t, a characteristic COD, δ_c, and a shape of the descending branch of this relationship, here assumed to be linear. (The steep rising slope at the left is used in the numerical solution and has negligible effect on the results. Ideally, this stiffness would be infinite.) The area under the curve shown in Fig. 8 is the fracture energy of the material, G_{Ic}, which here has a value of 1.38 in-lbs. The relationship between the LEFM parameters K_{Ic} and G_{Ic} is well known, and here yields a value of $K_{Ic} = 2.2 \, \text{ksi} - \sqrt{\text{in}}$. The value of tensile strength used is an estimate, mentioned above, based on the actual average measured value of splitting tensile strength of 620 psi [13]. The shape of the descending branch was chosen for simplicity, although it is known that the shape can have significant influence on process zone development [9]. The value of toughness is representative of relatively high strength concrete with large aggregate size (maximum aggregate size = 1.0 inch [3]).

3.2. Numerical methods

CRACKER [14] was used for the analysis. This code can be used to model either static or dynamic crack propagation with either LEFM assumptions or a nonlinear stress-vs-COD relation on the crack face. Through the use of interactive computer graphics, cracks can be initiated at any desired location during an analysis and can then propagate in directions not specified *a priori*. The cracks are modeled discretely as part of the mesh; the mesh changes as the cracks grow. Significant features of the implementation are;

- the use of explicit time integration for dynamic fracture;
- static solutions obtained either by dynamic relaxation or Crout elimination;
- triangular elements with quadratic shape functions to model the singularity at the crack tip;
- the integrated use of interactive computer graphics in all phases of the analysis.

In this investigation static solutions were obtained using dynamic relaxation [15]. This technique is a reliable approach to problems involving nonlinearities such as those encountered in the spring elements used to represent the nonlinear soil support of the pier. This explicit scheme does not require matrix inversion and no requirements are place on the order in which nodes or elements are numbered. This makes it ideal for problems in which the mesh is changing during the analysis.

Quadratic six-noded triangular elements are used because the singular stresses around the crack tip can easily be incorporated by shifting the side nodes to the quarter-point positions [16]. A second consideration is that automatic remeshing is generally easier using triangular elements.

Using the above approach, it is possible to obtain the solution for displacements and stresses at any time in the analysis. For both static and dynamic propagation, the asymptotic displacement solution around the crack tip is known in terms of the stress intensities, position relative to the crack tip, and crack velocity. Since the finite element solution provides the displacements around the crack tip, it is simple to solve for the stress intensities at any time using displacement correlation [17].

Once the stress intensities have been obtained, it is necessary to formulate propagation criteria to predict crack growth direction. In the present calculations, we use the maximum circumferential stress criterion to predict the direction of crack growth [18]. To use this criterion, the asymptotic solution is used to evaluate the direction of largest circumferential stress around the crack tip. The crack is then moved in that direction for the next time step. This is equivalent to moving in the direction that minimizes the mode II stress intensity.

For a static analysis, remeshing is performed after each static solution when the tip is moved a specified increment. Only local remeshing is performed around the tip. First, elements around the crack tip are deleted. A rosette of elements around the crack tip is then introduced and the annulus between the rosette and mesh is filled. We use the modified Suharra-Fukuda algorithm [19] to generate the new elements in this region. With the new mesh it is now possible to obtain a new solution and proceed to again propagate the crack.

The calculations described in this paper used both LEFM and a stress-vs-COD relation to model nonlinear fracture. The LEFM situation incorporates the singularity at the crack tip by shifting the side nodes of the elements to the quarter-point. Propagation is then in the direction of maximum circumferential stress. For the nonlinear analysis we also employ singular elements at the crack tip. This is to allow us to determine the direction in which the crack is to propagate. Indirectly, this also allows us to determine the point of instability of the nonlinear crack. In the analysis we apply a constant load. We then propagate the crack and can determine the stress intensity as a function of crack length. If the stress intensity remains small (say 1/100 of the critical stress intensity) then the nonlinear crack is stable and the nonlinear solution is valid. However, if this stress intensity becomes large, this implies that the nonlinear process zone is not supporting the load and that LEFM might, now be usable. This also implies instability of the crack and dynamic crack propagation.

3.3. Description of model

A number of meshes, differing mostly in detail around point A, were used in the analysis. These allowed us to investigate mesh sensitivity on our result. The overall coarsest mesh and loads are illustrated in Fig. 9. The re-entrant corner details of the finest mesh is shown in Fig. 10. Material properties for the concrete were:

$$E = 3.5 \times 10^6 \,\text{psi},$$

$$v = 0.10,$$

$$\rho = 0.086 \,\text{lb/in}^3.$$

Loads on the pier included the weight of the supported span, shown as column loads in Fig. 9, and self-weight. The pier support was modeled using a soil stiffness of $2.5 \times 10^6 \,\text{lb/ft}^3$. This is a reasonable value within the range of measurements reported in [3]. The soil reaction was simulated with nonlinear interface elements depicted in Fig. 9. In our analyses, the soil was not allowed to carry tensile stress.

4. Results

4.1. Crack initiation at point A: mesh dependency

Basing crack initiation on a stress calculated at or near a re-entrant corner is inherently problematical. This is shown quantitatively in Figs. 11–12. Figure 11 is the displaced shape for case 3, and Fig. 12 is the corresponding distribution of vectors of tensile principal stress. Concentration of tensile stress near point A is evident. However, the magnitude of principal stress at or near this point cannot be computed objectively. Figure 13 shows plots of stress along the line shown in Fig. 10 for the coarsest and an intermediate mesh. These are plots of unsmoothed results taken directly from each mesh. The trend towards increasing stress with

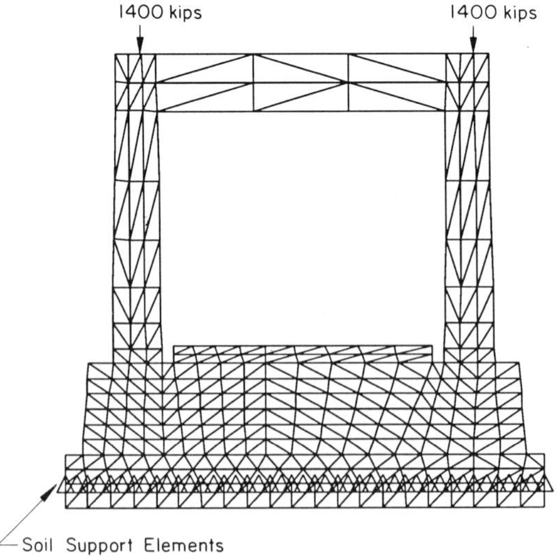

Fig. 9. Coarsest mesh used in present analyses. All elements are of quadratic order. Looking East.

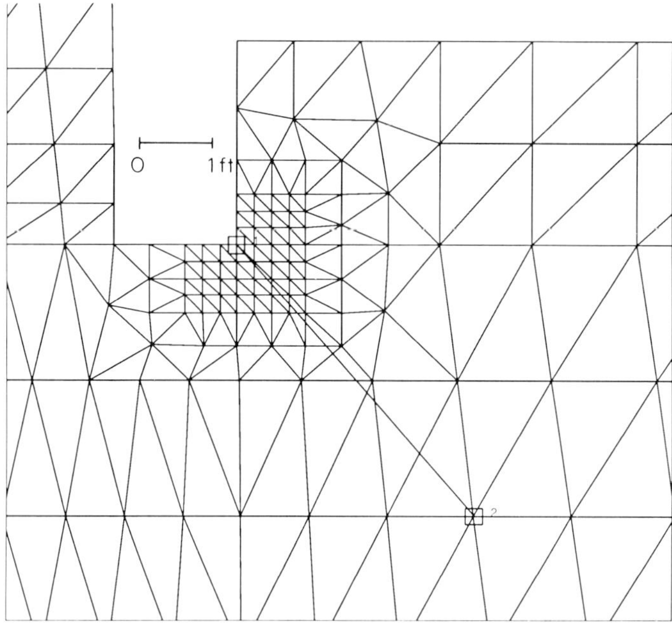

Fig. 10. Detail of finest mesh near Point A. Line 1–2 used in Fig. 13.

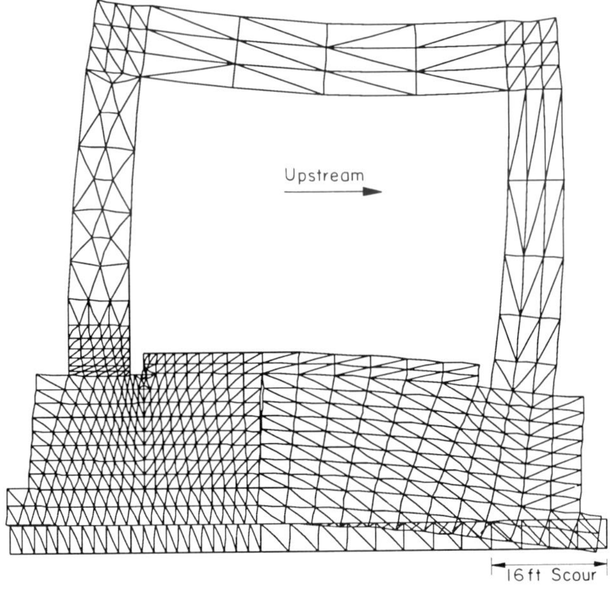

Fig. 11. Predicted displaced shape after 16 feet of scour. Magnification factor is 585. Looking East.

mesh refinement is evident. It is pointless to base crack initiation on the stress computed at point A.

It is just as pointless to turn immediately to LEFM to resolve this dilemma. What size crack should be tried at point A? How can one know *a priori* that LEFM is applicable for the size

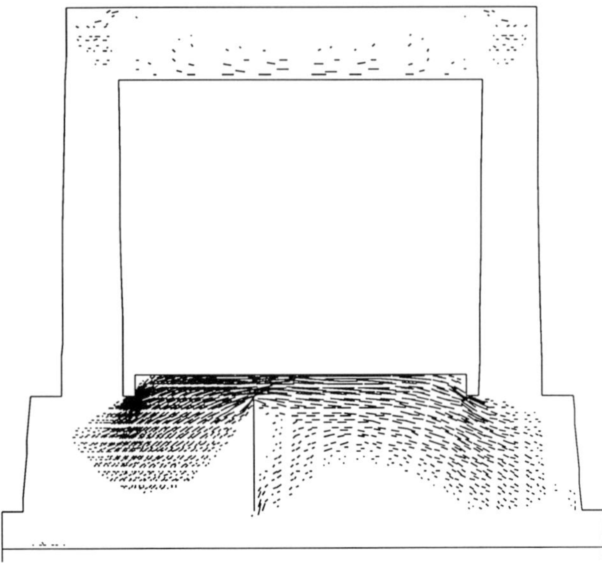

Fig. 12. Predicted distribution of vectors of tensile principal stress after 16 feet of scour.

selected? Rather, it is necessary to model the *process* of crack initiation at this point with the nonlinear fracture mechanics methodology described in Section 3.1 and in the following.

4.2. Crack initiation at point A: Nonlinear fracture mechanics

The next step in our analysis was to study the *stability* of process zone development around point A. This was done in the following manner:

1. Predict the likely direction of process zone formation starting from point A. The direction was assumed to be the curve normal to the principal tensile stress direction in the near field, computed from an elastic analysis.
2. Insert the process zone constitutive model shown in Fig. 7 along this direction, initially for a distance of a few feet starting from point A.
3. Perform nonlinear analyses for increasing lengths of scour, with a refined mesh in the suspected process zone region. These analyses will indicate what length is necessary to initiate process zone formation as discussed in Section 4.2.1.
4. Determine the length of scour required to initiate unstable crack propagation. This procedure is discussed in Section 4.2.2.

4.2.1. Process zone initiation
The discussion in Section 4.1 should make it clear that a process zone would theoretically *begin* to form at point A in the limit of mesh refinement and at the limit of zero lead. Two questions which then arise are:

1. What level of practical mesh refinement will initiate process zone formation at an actual level of load?

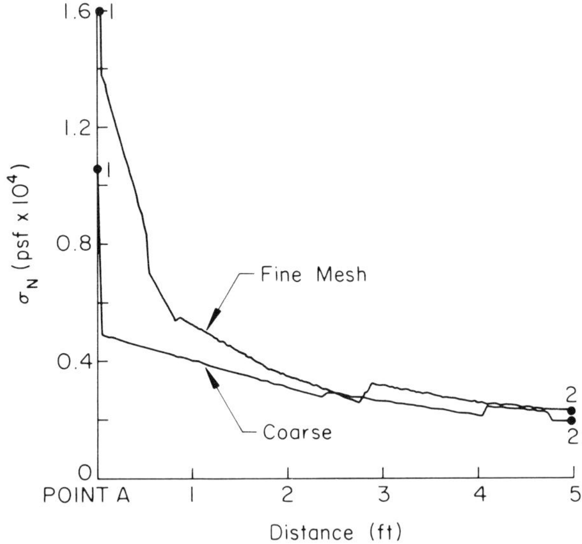

Fig. 13. Stress normal to line shown in Fig. 10. After 16 feet of scour.

2. Does mesh refinement beyond this level influence subsequent results?

Figures 14 and 15 address those questions.

Figure 14 shows that, at 24 feet of scour and at the level of mesh refinement shown in Fig. 10, no process zone has begun to form. This is because this level of discretization is insufficient to generate a stress at Point A greater than the tensile strength. In contrast, Fig. 14 also shows the beginning of process zone formation with eight more feet of scour. The observation here is that in the presence of very high stress concentrations sufficient refinement is necessary to *initiate the damage process in the model.*

It must be noted that for refinement less than that shown in Fig. 14, an incorrect conclusion would be drawn: no cracking would be predicted at 32 feet of scour.* At any level of mesh refinement greater than or equal to that shown process zone initiation would be predicted, but the length of process zone developed by the model would depend somewhat on the mesh. However, since the singularity at point A is now gone, the mesh dependence would be of the same type which occurs in any finite element analysis without singularities, but with a local stress gradient: the analysis is objective, and will converge in the usual *h*- or *p*-sense.

4.2.2. *Process zone stability*
The initiation of a process zone does not necessarily signal unstable propagation in general, and it did not in this case. Figure 14 also shows a fully converged result from 32 feet of scour. A process zone of approximately 4 inches has formed, and no true cracking has occurred. The situation remains much the same at 36 feet of scour. This is significantly more than the 28 feet

---

*The shape of the spikes in the stress plots shown in Figs. 14 and 16 are the result of using quarter-point elements around the end of the process zone. These elements can produce stresses which are constant and which vary as $\sqrt{r}$, where r is the distance from the tip of the process zone. The magnitude of the spikes is the result of using a non-infinite pre-peak stiffness in the process zone model shown in Fig. 8.

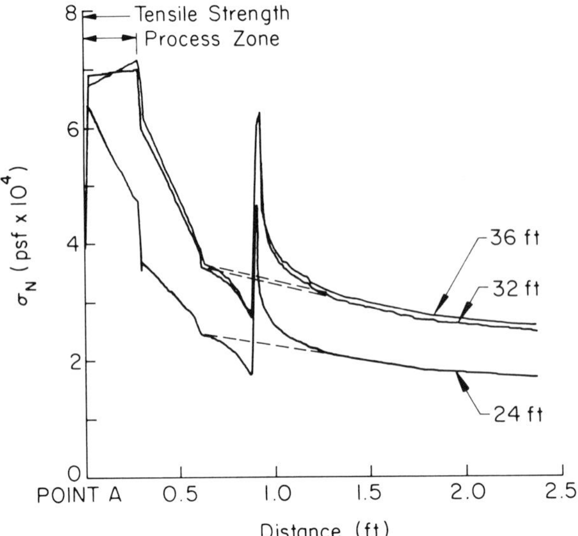

Fig. 14. Stress normal to line shown in Fig. 10 for various lengths of scour.

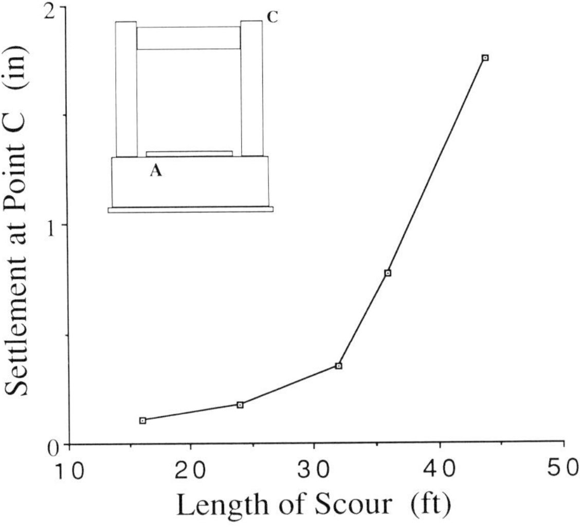

Fig. 15. Predicted settlement at top of upstream (south) column versus length of scour.

assumed in [3] to have been sufficient to trigger unstable crack propagation. We investigated a number of other possible contributions to crack instability: complete failure of the tie beam, discrete modeling of all shear studs in the plinth reinforcements, and even halving of the tensile strength. None of these conditions triggered process zone instability for scour lengths up to 36 feet. Additional scour might cause instability, but this alone did not appear to be the complete answer. Another process, described in the following section, must have occurred to raise the level of tensile stress in the area surrounding the developing process zone to trigger its complete development.

4.3. Crack propagation

The result shown in Fig. 14 suggested increasing the length of scour. However, an additional problem then arises. At a scour length of 41 feet, and with equal column loads and with no tension allowed in the soil the pier becomes kinematically unstable. This is a failure mode that is unrealistic, since cracking, rather than overturning, did occur, as shown in Fig. 4. How can additional scour beyond 41 feet occur without this inadmissable failure mode?

The answer to this question, and a key point in understanding the subtle and complex sequence of events leading to ultimate failure, begins to become clear through Fig. 15. It shows the settlement that is predicted beneath the bearings supporting the south girders of bridge deck sections 3 and 4 on pier three. Since each deck section is supported at four such points, and the deck has finite torsional rigidity, settlement at this point leads to unloading of the south column of the pier. The load would then shift to the other participating columns, but mostly to the north column of pier three. This shift changes the kinematic stability situation of the pier, and allows scour to proceed beyond the mid-point of the plinth. *This is the additional process which we believe played an importaat role in the collapse process.*

We did not perform an analysis of the deck/piers system. Rather, we assumed that at 44 feet of scour one-half of the load in the south column would have shifted to the north column, and proceeded with another nonlinerar analysis. Direct evidence that at least 44 feet of scour occurred before failure of the plinth is shown in Fig. 4. The results are shown in Figs. 16 through 19.

Figure 16 shows that process zone formation has been completed over a distance of about 4 inches and that the length of the crack is now about 14 inches. However, although this is a fully converged, equilibrium result, it does not satisfy completely the constitutive model in the following way. The stress intensity factor computed at the tip of the process zone should

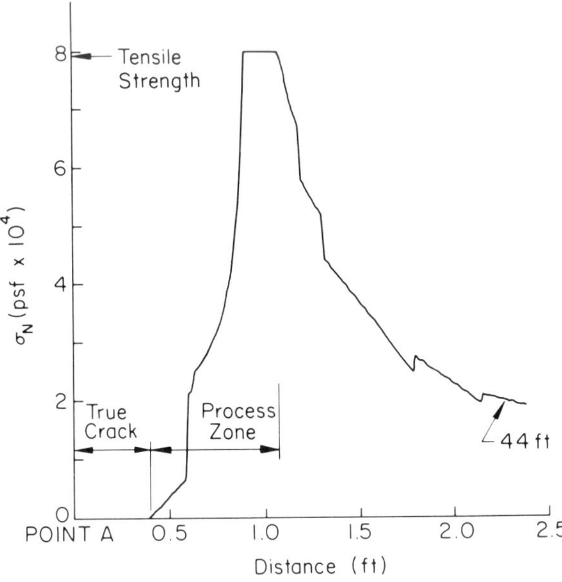

Fig. 16. Stress normal to line shown in Fig. 10 for 44 feet of scour and column load shift.

theoretically be zero, or practically less than 1/100th of K_{Ic}. Here its value is $1.8\,\text{ksi}\,\sqrt{\text{in}}$ $(75\,\text{ksf}\,\sqrt{\text{ft}}.)$, or about 80 percent of K_{Ic}. This means that the length of the crack, true plus process zone, caused by 44 feet of scour should be much longer than that shown in Fig. 16.

At this point it is apparent that a transition from nonlinear to LEFM is in process. Figure 17 shows the stress intensity factor history for crack propagation under 44 feet of scour, assuming LEFM. It shows that for the initial crack length considered, about 12 inches, K_I is already higher than K_{Ic} and that unstable crack propagation would occur.

In contrast, for 24 feet of scour a crack nearly 5 feet in length would have been required to precipitate instability. Figure 18 is a schematic of the transition process. Clearly, LEFM has no applicability in this problem for initiation or for crack growth less than about 12 inches. LEFM would become applicable when the length of process zone became small, say less than a tenth of the total crack length. Figure 16 showed that the process zone length was about 8 inches when total crack length was about 13 inches. However, in [9] it was shown that the process zone length is not constant, tends to be longest during the initiation phase, and will decrease if the stress parallel to the direction of propagation is compressive which is the case here. These observations together suggest that, with a representative process zone length of about 6 inches, LEFM would become applicable at about 5 feet of cracking.

The final trajectory, predicted by mixed-mode LEFM and corresponding to the K_I history shown in Fig. 17, is presented in Fig. 19. The analysis was stopped after about 20 feet of cracking, with the crack tip approaching the bottom of the pre-existing crack. The pre-existing crack can be seen also to be opening. Agreement with the observed trajectory shown in Fig. 4a, including interaction with the pre-existing crack, is very good.

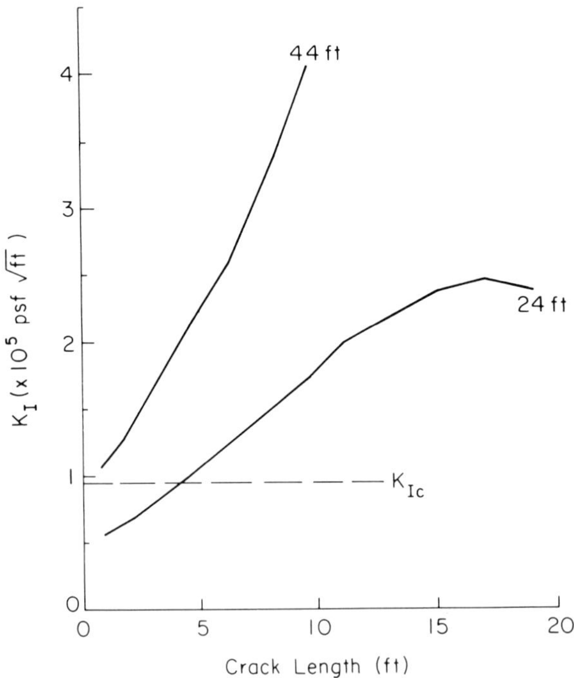

Fig. 17. K_I versus crack length. The trajectory for 44 feet of scour is shown in Fig. 19. $(1000\,\text{psf}\,\sqrt{\text{ft}} = 24\,\text{psi}\,\sqrt{\text{in}})$

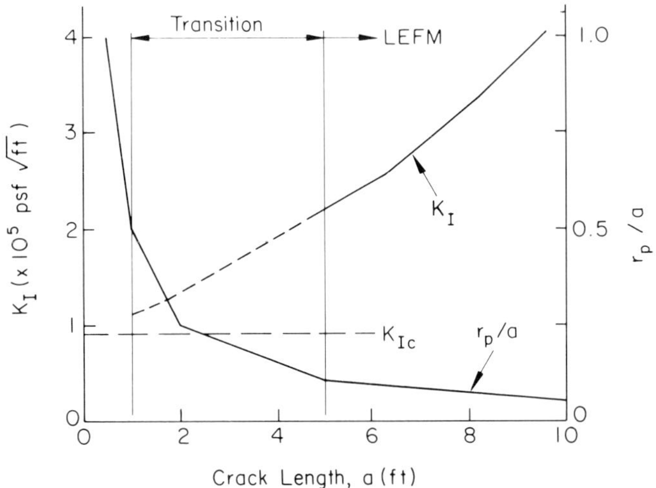

Fig. 18. Schematic of transition from nonlinear to LEFM. Process zone length is r_p.

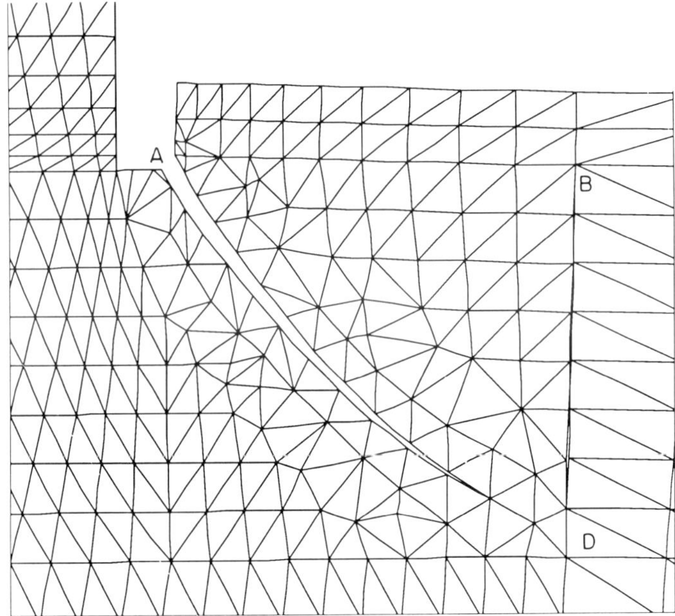

Fig. 19. Predicted trajectory with 44 feet of scour. Pre-existing crack runs from points B to D. Magnification factor is 500.

5. Discussion

As shown here, the events leading to collapse were considerably more complex than previously thought. Markedly more scour and a redistribution of column loads has to occur to generate conditions not just for initiation but for unstable propagation of the fatal crack.

By using nonlinear fracture mechanics the problems associated with predicting initiation in a region of high stress gradient using finite elements can be avoided. It was not necessary to call on size-effect to lower strength, or to require substantial tension in the soil to increase stress. For this structure, unstable process zone formation lead quickly to a transition to crack propagation under the rules of LEFM. No crack arresting mechanism existed.

The implications of this failure analysis for many other bridges of this type are serious. Although the unreinforced plinth was miraculously able to sustain scour over more than half its length, the bridge itself exhibited little effective ductility: only a few inches of settlement of the upstream column, an amount probably not readily noticeable, occurs before unstable cracking causes kinematic instability of the deck sections of spans 3 and 4. Prevention of scour as suggested in [3] is certainly a first line of defense, but one more subject to the whims of nature than sound structural design.

It is strongly recommended that existing bridges of this design be retrofitted with crack arresting mechanisms. Here again nonlinear and LEFM can be put to effective use in computing the amount and distribution of post-tensioned steel in the plinth to perform the necessary arrest function.

6. Conclusions

Based on the linear and nonlinear fracture mechanics analyses reported here, we can conclude:

1. Although the proximal cause of the collapse of the Schoharie Creek Bridge was scour beneath the plinth of pier three, a *necessary* secondary cause was unstable crack propagation in that plinth.
2. It is unlikely that *initiation* of the critical crack in that plinth began until about 28 feet of scour had occurred.
3. At about that length of scour, stable process zone, not true crack, initiation began. This is a critical refinement of the conclusion in [3] that crack initiation *and* unstable propagation occurred at about this length of scour.
4. Fracture mechanics and its implied size effect, although laudatory and of historical importance in their use, were mis-interpreted in [3]. Size effect does not reduce the local tensile strength: it reduces the apparent modulus of rupture. It is the former, and not the latter, which controls crack initiation at a particular point in the plinth.
5. With lengthening of the scour from 28 feet through 36 feet, the process zone grows only slightly and no true cracking is predicted, even if the assumed toughness is reduced by a factor of two, the shear studs are represented discretely, or the tie beam fails completely.
6. Under the action of equal column loads, the pier becomes kinematically unstable at 41 feet of scour and with no soil tension permitted.
7. Scour causes settlement which in turn causes a redistribution of column loads. Unloading of the upstream column of pier three allowed scour to increase without causing kinematic instability and without requiring tensile stress in the soil.
8. Process zone formation became unstable at about 44 feet of scour. By this time, true cracking has begun with an initial process zone length of about 8 inches.

9. The transition from process zone formation to LEFM occurs within about the first 5 feet of cracking, and crack propagation was unstable through the transition to at least a crack length of 20 feet.

10. The length of scour at crack instability and the predicted trajectory of the crack were in very good agreement with observations.

11. Unstable cracking causes complete loss of support from the upstream column of pier three. This causes kinematic instability of the deck sections of the spans 3 and 4.

12. The unreinforced *plinth* shows remarkable strength in supporting scour over more than half its length. However, the *bridge* has relatively little effective ductility since only a few inches of column settlement occur before unstable crack propagation precipitated collapse.

13. Although scour prevention is desirable [3], it is less reliable than proper retrofitting of many existing bridges of this type. It is strongly recommended that plinths be equipped with crack arrest capability. Fracture mechanics should be used to design safe and cost-effective post-tensioning schemes for this purpose.

Acknowledgements

This work was partially supported by National Science Foundation Grant PYI-835 1914. Simulations were performed in Kansas State's Center for Research in Computer Control and Automation, and in Cornell's Computer-Aided Design Instructional Facility and Program of Computer Graphics.

References

1. Interim Report, 'Collapse of the Thruway Bridge,' New York State Disaster Preparedness Commissions, Albany, New York, May 27, 1987.

2. Preliminary Report, 'Collapse of the Thruway Bridge at the Schoharie Creek,' Wiss, Janney, Elstner Associates, Inc., and Mueser Rutledge Consulting Engineers, New York State Thruway Authority, Albany, New York, June 15, 1987.

3. Final Report, 'Collapse of the Thruway Bridge at Schoharie Creek,' Wiss, Janney Elstner Associates, Inc., and Mueser Rutledge Consulting Engineers, New York State Thruway Authority, Albany, New York, November, 1987.

4. Z.P. Bažant, J. Kim and P. Pfeiffer, in *Application of Fracture Mechanics to Cementitious Composites*, S. Shah (ed.), Martinus Nijhoff Publishers (1985) 197–246.

5. P.F. Walsh, *Indian Concrete Journal* 46, No. 11 (1972).

6. A. Hillerborg, *International Journal of Fracture* 51 (1991) 95–102.

7. P.E. Petersson, 'Crack Growth and Development of Fracture Zones in Plain Concrete and Similar Materials,' Rept. TVBM-1006, Division of Building Materials, Lund Institute of Technology, Sweden, 1981.

8. A. Hillerborg, M. Modeer, and P.E. Petersson, *Cement and Concrete Research* 6 (1976) 773–782.

9. A.R. Ingraffea and W.H. Gerstle, *Application of Fracture Mechanics to Cementitious Composites*, S. Shah (ed.), Martinus Nijhoff Publishers (1985) 247–286.

10 T.J. Boone, P.A. Wawrzynek and A.R. Ingraffea, *International Journal of Mechanics, Mining Science and Geomechanics Abstracts* 23, No. 3 (1986) 255–265.

11. *Fracture Toughness and Fracture Energy of Concrete*, F.H. Wittman (ed.), Elsevier Publishers (1986).

12. *Fracture Mechanics of Concrete: Structural Application and Numerical Calculation*, G. Sih and A.D. Tommaso (eds.) Martinus Nijhoff Publishers (1984).

13. A.M. Neville, *Properties of Concrete*, 3rd. Edition, Pitman Publishing, London (1981).

14. D.V. Swenson and A.R. Ingraffea, *Computational Mechanics* 3 (1986) 381–397.

15. P. Underwood, in *Computational Mechanics for Transient Analysis*, North Holland, Amsterdam (1983) Chapter 5.

16. R. Barsoum, *International Journal of Numerical Methods in Engineering* 11 (1977) 85–98.
17. C.F. Shih, de Lorenzi and M.D. German, *International Journal of Fracture* 12 (1976) 647–651.
18. F. Erdogan and G.C. Sih, *ASME Journal of Basic Engineering* 85 (1963) 519–527.
19. R.D. Shaw and R.G. Pitchen, *International Journal for Numerical Methods in Engineering* 12 (1979) 93–99.

PART 3. NONLINEAR FRACTURE OR DAMAGE MODELS AND SIZE EFFECT

International Journal of Fracture **51**: 95–102, 1991.
Z.P. Bažant (ed.), Current Trends in Concrete Fracture Research.
© 1991 *Kluwer Academic Publishers. Printed in the Netherlands.*

Application of the fictitious crack model to different types of materials

ARNE HILLERBORG
Lund Institute of Technology, Lund, Sweden

Received 1 June 1990; accepted 1 November 1990

Abstract. The fictitious crack model is a non-linear fracture mechanics representation which has been well known and generally accepted for application to concrete fracture for about 10 years. It is a general model which in principle is applicable to all kinds of materials. At present, however, practical and numerical difficulties prevent its application to many types of materials, particularly to metals. In the future, better knowledge of material properties and more powerful computer programs may widen the potential field of application. The possibility of application to different types of materials is discussed.

1. Introduction

The basic ideas behind what has later become known as the fictitious crack model (later introduced alternative – and maybe better – names are cohesive crack model [2], cohesive zone model or damage zone model [4]), were first published in [1]. Normally it is illustrated as in Fig. 1, showing a tensile test in deformation control. In such a test, a fracture zone is formed before final fracture. The formation of the fracture zone involves damage of the material in the zone, which results in a decrease in stress transfer. This decrease in stress causes unloading of the material outside the fracture zone, with a corresponding decrease in strain. This strain decrease and the increasing total deformation have to be taken up as an additional deformation w within the fracture zone. This brings us to the material description which was introduced in the fictitious crack model.

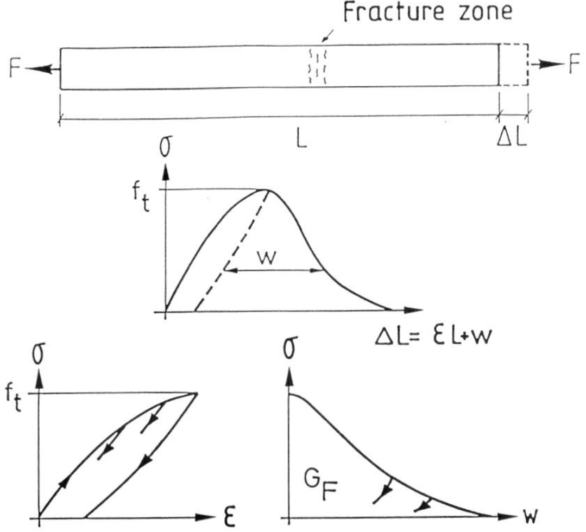

Fig. 1. The basis of the fictitious crack model, demonstrated in a tensile test. The area G_F is called fracture energy.

The complete stress-deformation properties of a material can be described by means of two relations, one stress-strain relation ($\sigma - \varepsilon$), including unloading branches, which is valid for all the material, and one stress-displacement relation ($\sigma - w$), also including unloading branches, which gives the additional deformation due to a fracture zone.

Although the model is mostly illustrated by means of a tensile test, it may also be used as a general description of deformation properties where tensile fracture zones appear, i.e. where the stresses start to decrease under the action of an increasing deformation, often referred to as the descending branch of a stress-deformation (or stress-strain) relation. For concrete and some other materials, the two curves from a tension test may be assumed to be material properties in all those cases where fracture is dominated by tensile stresses. In many other situations, the curves to be applied depend on the stress situation. Thus the shape of the stress-strain curve may be strongly influenced by the three-dimensional stress state, which is the case with metals, for example.

The application of the model is illustrated in Fig. 2, which shows the formation and development of a tensile fracture zone in a bent beam. The fracture zone starts forming when the tensile stress reaches the strength f_t. The fracture zone (the fictitious crack) propagates in such a way that the stress at the tip of the fictitious crack always has the value f_t. Within the fictitious crack, the stress depends on the additional deformation w within the fracture zone (the width of the fictitious crack). Outside the fictitious crack, the stress-strain curve is valid.

The additional deformation w takes place within a fracture zone of a certain width (length in the stress direction). In the numerical application, some assumption has to be made regarding this width. In the earliest applications by means of the finite element method, the additional deformation within the fracture zone was modelled as a separation of elements a distance w, introducing stresses σ between the separated elements, according to the $\sigma - w$ relation. This

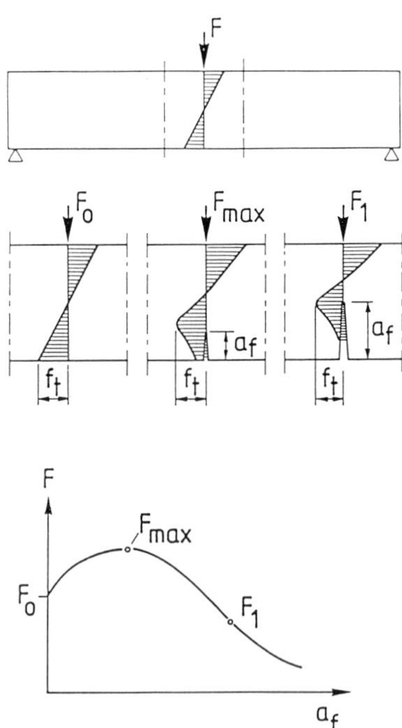

Fig. 2. Formation and growth of a fracture zone and a real crack in a bent beam.

stress transfer between the separated elements corresponds to a stress transfer across a crack. As a real crack cannot transfer such stresses, the assumed stress-transferring crack was referred to as a fictitious crack – hence the name of the model.

The term fictitious crack model thus originates from the finite element application of the model, which, in itself, does not assume any fictitious crack or infinitely narrow fracture zone, but just states that the additional deformation within the fracture zone has to be taken into account as a stress-displacement relation.

One consequence of the model is that a complete stress-strain curve, including a descending branch after the peak, does not exist as a material property. In cases where such curves are shown, they have to be referred to a certain gauge length, over which the additional deformation within the fracture zone has been averaged.

In most numerical applications, the stress-strain relation has been assumed to be linear, simply because any other assumption complicates the analysis very much. It must, however, be emphasized that the model is not limited to the linear stress-strain relation. With many materials, it is necessary to take a non-linear relation into account, and consequently the model has not yet been applied to these materials. In the future, more powerful computers and improved finite element programs will certainly make such applications possible.

One advantage of the fictitious crack model over conventional fracture mechanics is that in addition to being used for analysing the stability and growth of cracks, it can also be used to analyse the formation of cracks; see the example in Fig. 2.

2. The influence of a three-dimensional stress state

Stress-strain relations for isotropic materials

For many materials, like metals, the stress-strain properties in one direction are strongly dependent on the stresses and strains in the perpendicular directions. Thus, for example, the fact that the ultimate stress in triaxial tension may be much higher than in uniaxial tension has to be taken into account in the analyses. Before this can be done, however, a sufficiently correct description has to be available regarding the properties under triaxial states of stress. The numerical method must also be able to deal with non-linear stress-strain relations.

According to the fictitious crack model, the fracture zone starts developing when the first principal stress has reached a peak value. The stress-strain relations that are attained before the peak stress is reached are often referred to as the pre-peak properties.

There is a distinct difference in pre-peak properties between what can be referred to as microcracking materials and yielding materials.

In a microcracking material, the non-linearity of a stress-strain relation depends on the opening of microcracks perpendicular to the first principal stress. These microcracks can open without causing any major lateral strains or stresses. Therefore the material properties in the tensile direction can often be assumed to be the same as in uniaxial tension without risking any major errors. This simplifies the numerical analyses considerably, and it is probably the only assumption which has been used in applications so far. The assumption is only acceptable as long as one tensile stress dominates the fracture, whereas the stresses in the perpendicular directions are far below the corresponding strengths.

In a yielding material, on the other hand, the non-linearity of a stress-strain relation depends on plastic deformations, which are always accompanied by lateral strains or stresses, as the corresponding volume changes are minor. If the lateral strains can develop unhindered, they do not give rise to any lateral stresses. This is called a plane stress condition. The opposite case, i.e. plane strain condition, occurs when the lateral strain development is totally prevented, e.g. by the surrounding material. In this case large lateral stresses are developed and, in their turn, give rise to a triaxial stress state, which has a great influence on the response of the material, e.g. increased yield strength and ultimate strength, and also a decreased strain at peak stress.

Pure plane stress and plane strain are theoretical limiting cases, which hardly ever occur in practice, where the situation is instead somewhere between these limits, because the strains are only partly counteracted by the surrounding material. A well known example is a through crack in a metal plate. When the crack starts growing, the material at the crack front can be assumed to be under plane stress conditions at the ends of the crack, close to the plate surface, but under plane strain conditions at the center of the crack. Along the crack front, there is some kind of transition between these two conditions.

The transition between plane stress and plane strain conditions makes the numerical analysis of fracture in yielding materials extremely complicated. The marked non-linearities of the pre-peak properties of such materials and the large deformations associated with yielding are additional complications.

Stress-displacement relations for isotropic materials

The development of a tensile fracture zone is always accompanied by the formation of micro-cracks or micro-voids. These cracks or voids decrease the possible stress transfer across the fracture zone. As the cracks or voids grow, the displacement increases, whereas the transferred stress decreases. The displacement in the fracture zone can take place without giving rise to lateral strains, as the volume may increase when the cracks or voids form. The situation within the fracture zone may therefore be assumed to correspond to plane stress conditions. One consequence of this is that the stress-displacement relation may be assumed to be a material function, as long as the displacement direction is unchanged during the fracture process. The relation may depend on the stresses and strains in the perpendicular directions.

If the displacement direction changes during the fracture process, the situation becomes more complicated. One possibility is that the fracture zone starts as a pure tensile fracture, but that shear stresses and displacements develop later within this zone. Another possibility is that a new fracture zone starts forming in a skew direction to the first tensile fracture zone.

Non-isotropic materials

In a non-isotropic material, the tensile fracture zone may form in a skew direction to the first principal stress. In such a fracture zone shear stresses also occur and have to be transferred. The criterion for the formation of a fracture zone will also depend on shear stresses. This, of course, introduces new practical problems for determining the material properties and for the numerical application.

3. Applications

Ordinary concrete

The fictitious crack model has so far mainly been applied to ordinary concrete, for which it was first developed. It has proven very suitable for this purpose, as concrete is a micro-cracking material with a stress-strain curve which may, as a rule, be assumed to be linear all the way up to the peak.

A typical stress-displacement relation for a tensile fracture zone in concrete is shown in Fig. 3.

The size of the fracture process zone for concrete is often of the order of 100 mm at maximum load on a structure, which makes linear elastic fracture mechanics (LEFM) unsuitable for normal structural sizes. The possibility to analyse not only the propagation of cracks, but also the formation of cracks in an uncracked structure, also makes the fictitious rack model superior to LEFM for practical application to fracture analyses of concrete structures.

The applicability of the model has been confirmed by comparisons with the results of tests with different types of laboratory specimens. The model has also been used for analysing the behavior of different types of concrete structures; see e.g. [2].

Fiber reinforced concrete

In fiber reinforced concrete, the fibers bridge the cracks and transfer stresses. The fictitious crack model is therefore well suited for fiber-reinforced concrete. Examples of applications can be found in [3].

If concrete is heavily reinforced with fibres, this will change the pre-peak behaviour of the material, so that the stress-strain relation becomes markedly non-linear. In such cases, a linear approximation of the stress-strain curve is not acceptable, which complicates the analysis.

Rock

Rock has properties which are similar to those of concrete. The fictitious crack model may thus be applied to rock as well.

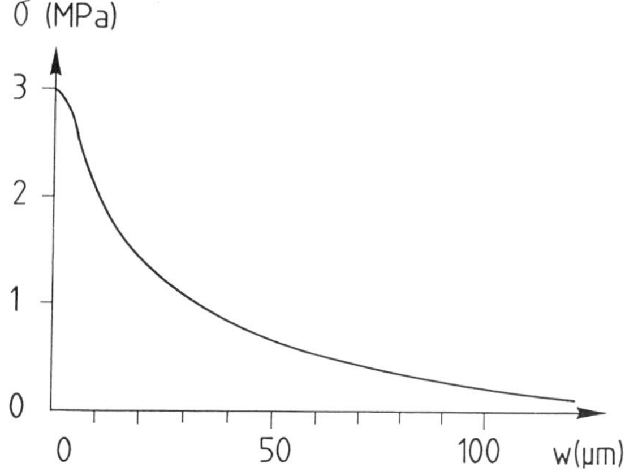

Fig. 3. A typical stress-displacement relation for concrete.

100   *A. Hillerborg*

Fiber reinforced plastics

Fiber reinforced plastics, e.g. carbon fiber reinforced epoxy or glass fiber reinforced polyester, are also micro-cracking materials, with cracks forming in the matrix before the peak stress is reached. Therefore the fictitious crack model may be well suited for application to such materials.

Just like fiber reinforced concrete, fiber reinforced plastics may also display a non-linearity of the stress-strain curve which is too large to make a linear approximation acceptable. In many cases a linear approximation seems acceptable, however.

In fiber reinforced plastic sheet material, the fibers are often oriented so as to make the material strongly non-isotropic in the plane of the sheet. Naturally this has to be taken into account in the application of the model.

The fictitious crack model has been applied to fiber reinforced plastics by Bäcklund and coworkers under the name 'damage zone model' [4]. It has been shown to have certain advantages over earlier models for analysing the formation of cracks from holes.

Wood

The fictitious crack model can also be applied to wood, which may be looked upon as a micro-cracking material. Wood is such a non-isotropic material that the fracture zones normally form along the grains, almost independently of the stress direction.

One example of a practical application is the analysis of the formation of shrinkage cracks due to drying [5]. Among other things this analysis shows how the size of the wood specimen influences cracking. It is well known from practice that small pieces of wood do not crack due to drying, whereas large ones do. The reason for this can easily be demonstrated by means of the fictitious crack model.

Metals

Most metals are yielding materials, which means that the triaxiality of stresses and strains causes great difficulties for the numerical application of the fictitious crack model. It will therefore take a long time before the finite element programs and the knowledge of the material properties are sufficiently developed to perform such analyses. Some qualitative conclusions can, however, be drawn from the model.

A cleavage fracture, i.e. a tensile fracture resulting in the separation of material along a plane which is perpendicular to the first principal stress, does not develop until the local tensile stress is much higher than the tensile strength determined in a standard test. In order to build up such high tensile stresses in one direction, it is necessary to have high tensile stresses also in the perpendicular directions; otherwise the material will yield instead of developing a cleavage fracture. A cleavage fracture cannot occur where the material is close to plane stress conditions, e.g. close to a surface. Here the fracture will instead be due to excessive shear yielding deformations at about 45° to the tensile stress direction. This type of fracture near a surface is sometimes referred to as a shear lip.

Further away from the surface, the perpendicular stresses may be large enough to make the first principal stress reach the level where a tensile fracture zone starts forming, resulting in a cleavage fracture.

Another case which can be discussed and possibly also numerically analysed with the model is the fracture of a round bar in a tension test; see Fig. 4. At the top of the figure, hypothetical material properties are shown. At maximum load, corresponding to the stress f_1, the material somewhere along the bar starts to develop large plastic deformations, and at this section failure will occur after a fracture zone has developed. The large plastic deformations give rise to large lateral deformations, resulting in necking and a decreasing bar area. As the deformations increase, the lateral stresses increase, also causing the stresses in the interior of the bar to increase due to the triaxial state of stress. The increasing mean stress partly compensates for the decreasing area. The highest tensile stresses appear at the center of the bar. When these stresses have reched the critical level f_2, a cleavage fracture starts (a fictitious crack). This will eventually lead to an unstable situation with the fracture zone rapidly spreading outwards, causing a sudden failure. At the surface of the bar, a shear lip may appear due to the limited lateral stresses. If a shear lip appears or not depends on the shape of the surface at the fracture stage, which, in its turn, depends on the material properties.

Not much is known regarding the material properties to be introduced in the application of the fictitious crack model to metals. In order to shed some light upon this situation, a test has been conducted on a steel bar which was turned to the shape shown in Fig. 5. This shape gives rise to high lateral stresses in the fracture zone and a cleavage fracture without much yielding. The descending branch of a stable tensile test may therefore at least give an idea of the stress-displacement curve to be used in a cleavage tensile fracture. The result is shown in the figure.

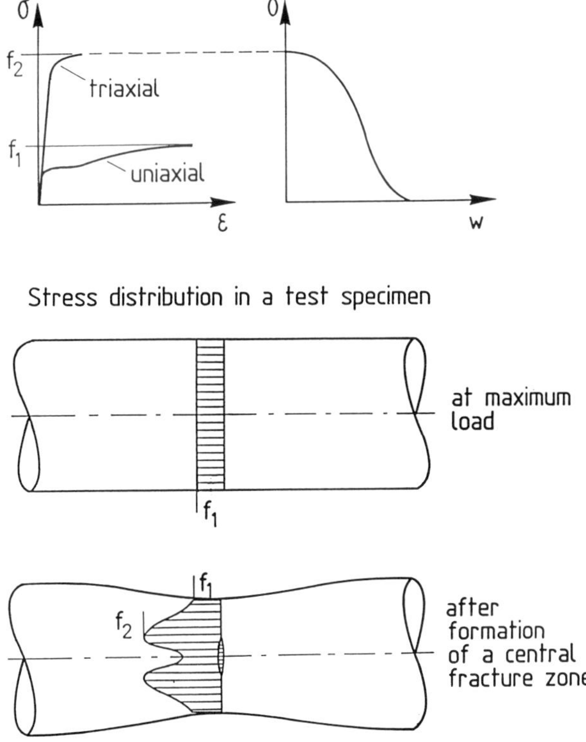

Fig. 4. Possible stress distributions in a tension test, where a cleavage fracture zone starts from the center.

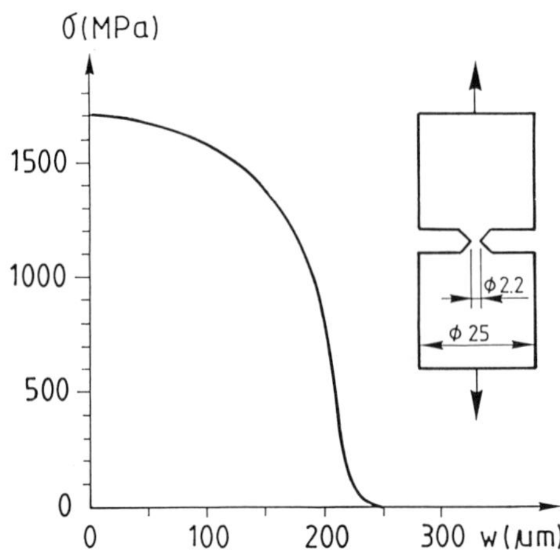

Fig. 5. A tension test to determine the stress-displacement curve in cleavage fracture of steel. The relation shown was measured for a mild steel with a yield strength of 300 MPa and an ultimate strength of 410 MPa.

4. Concluding remarks

The fictitious crack model is a general material model, which, in principle, could be applied to all types of tensile fracture if only the material properties were known and adequate tools for the numerical analyses were available.

So far, it has only been possible to apply the model to micro-cracking, non-yielding materials with an approximately linear stress-strain relation up to the peak stress. An example of such a material is concrete. The model has also been successfully applied to some types of fiber-reinforced concrete and plastics as well as to wood.

More fully developed numerical methods and more extensive knowledge of the material properties will certainly make it possible in the future to apply the model to other materials as well.

Even where a numerical application is not possible at present, some qualitative conclusions regarding fracture behaviour may be drawn by means of the model.

References

1. A. Hillerborg, M. Modéer and P.E. Petersson, *Cement and Concrete Research* 6 (1976) 773–782.
2. L. Elfgren (ed.), *Fracture Mechanics of Concrete Structures*, Chapman and Hall (1989).
3. A. Hillerborg, *International Journal of Cement Composites* 2 (1980) 177–184.
4. J. Bäcklund and C-G. Aronsson, *Journal of Composite Materials* 20 (1986) 259–302.
5. L. Boström, *Analysis of Shrinkage Cracks in Wood by means of Fracture Mechanics*, Division of Building Materials, Lund Institute of Technology, Report TVBM-3027 (1986).

International Journal of Fracture **51**: 103–120, 1991.
Z.P. Bažant (ed.), Current Trends in Concrete Fracture Research.
© 1991 Kluwer Academic Publishers. Printed in the Netherlands.

Features of mechanics of quasi-brittle crack propagation in concrete

YEOU-SHANG JENQ[1] and SURENDRA P. SHAH[2]

[1]Department of Civil Engineering, The Ohio State University, Columbus, Ohio 43210, USA;
[2]NSF Science and Technology Center for Advanced Cement-Based Materials, Northwestern University, Evanston, Illinois 60201, USA

Received 1 June 1990; accepted 1 November 1990

Abstract. Physical observations of crack propagation in concrete are discussed in the present paper. Based on these physical observations, different models proposed for mode I fracture of concrete are examined. In the past, verification of validity of fracture models has always been difficult. It is proposed that notch sensitivity and size effect be used as criteria to evaluate the validity of fracture models. In addition, a successful model should also satisfy a 'portability' condition, namely the model should be applicable to structures of an arbitrary geometry. Discussion on the effects of different loading conditions, i.e. mixed-mode loading conditions and different loading rates, on the fracture process in concrete is also presented.

1. Introduction

The fact that applications of linear elastic fracture mechanics (LEFM) have been shown to be successful for brittle materials, e.g. glass, has not resulted in its direct application to predict the quasi-brittle behaviour of concrete due to the heterogeneous properties of concrete. Since the basic compositions, microstructure, grain size, and atomic bonding properties of concrete are different from that of metal, the fracture behavior of concrete is, as expected, different from that of metals. Unlike metal in which the non-linear behavior (i.e. strain hardening and non-linear plasticity) mainly stems from the formation of dislocations, the major non-linear behavior of quasi-brittle materials results from the formation and branching of micro-cracks. Therefore, although non-linear fracture mechanics theories have been developed and successfully applied to metals [18], these theories may have to be modified for concrete applications.

In the present paper physical observations of crack propagation in concrete are discussed first. Based on these physical observations, different models proposed for mode I fracture of concrete are examined. In the past, verification of validity of fracture models has always been difficult. It is proposed that notch sensitivity and size effect be used as criteria to evaluate the validity of fracture models. In addition, a successful model should also satisfy a 'portability' condition; namely the model should be applicable to structures of an arbitrary geometry. Discussion on the effects of different loading conditions, i.e. mixed-mode loading conditions and different loading rates, on the fracture process in concrete is also presented.

2. Physical observations on quasi-brittle crack propagation

Since experimental results provide the foundation for theoretical formulation, it is necessary to look into the physical observations of quasi-brittle crack propagation in concrete reported by different laboratories. Some of the important physical observations can be summarized as follows:

a. Microcracking

The formation and propagation of micro-cracks in concrete are rather complicated. When a concrete structure is subjected to an external loading, microcracks will occur at the cement-aggregate interface and at the entrapped air voids. Existence of mirocracks in concrete has been reported by several investigators using microscopes, acoustic emission, and other methods [9, 10, 17, 27, 30, 31, 33, 35, 36, 37]. As the load starts to increase, some of the microcracks will grow and coalesce to form a macrocrack. However, some of the microcracks will be arrested by aggregates and air voids. To advance the cracks, a higher load and thus more energy is needed. As a result, crack branching may also occur at this stage. The coalesce and arrest processes of microcracks are complex and time-dependent, which is an important factor in the study of creep and dynamic effects on concrete behavior. When the applied load reaches the critical load, one of the macrocracks starts to propagate. If the test is controlled by displacement, the macrocrack will continue to grow and finally break the concrete structure into two halves. It should be noted that the formation of microcracks may also result from strain incompatibility between aggregates and hardened cement paste matrices.

b. Inclusion toughening

One of the toughening mechanisms in concrete is due to the arrest and/or branching of a matrix crack. Due to this deflection toughening effect, a larger surface area is generated and a higher load and energy are needed to continue to propagate this crack. This inclusion toughening mechanism has been studied by Maji and Shah [30] using a model concrete. Acoustic-emission (AE) signals produced during crack propagation were used to identify the location of the crack tip. Figure 1 gives the AE source locations of a model concrete at different loading stages. Assuming that the AE source location is the tip of a propagating crack, it can be seen that the first peak load was reached before the crack was arrested by the model aggregate (stages 1 and 2). The AE signals also show the existence of pre-peak stable crack growth. As the crack propagated further, it encountered a tougher model aggregate and its path was deflected. Due to the arrest and deflection mechanisms, the applied load has to be increased to propagate the crack (note stage 3 and 4 and the second peak in Fig. 1). Furthermore, based on the study of acoustic emission signals, different modes of fracture, i.e. tension and shear modes, were observed despite the fact that only a tensile load was applied [30]. It is believed that the shear mode was introduced due to pull-out of aggregates. It should also be pointed out that inclusion toughening is more important for regular strength concrete as compared to high strength concrete where the fracture path often goes through the aggregates [12].

c. Surface roughness

Due to the aforementioned inclusion toughening mechanism, cracks generally branch around aggregates. As a result, a crack randomly propagates in concrete and generally the crack surface is tortuous. The roughness of the crack surface is dependent on the relative toughness and size of the aggregates and properties of matrix and interface. The actual surface area of new cracks is likely to be larger than that determined by assuming a smooth fracture surface. To accurately calculate the surface energy of concrete based on the work-of-fracture method, a correct

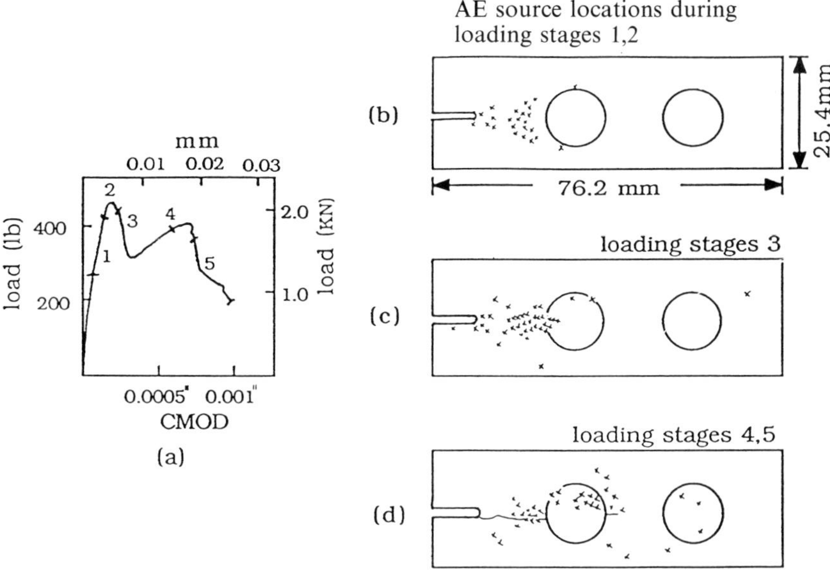

Fig. 1. Acoustic emission location at different loading stages.

measurement of actual fracture surface area is needed. Unfortunately, due to the difficulty in measuring the asperity of a crack and hidden fracture surface area, a smooth crack with zero asperity is often assumed. As a result, the specific energy dissipation (or fracture energy) calculated from the measured total input energy based on this assumption is only approximate and will be higher than that of actual surface energy.

d. *Three dimensional crack profile*

It was reported by Swartz and Refai [36, 37] that crack profile along the width of a concrete beam is not straight. More interestingly, unlike a channelized crack profile observed in a steel beam, a reverse channel shape of crack profile (Fig. 2) is observed. This reverse channel profile

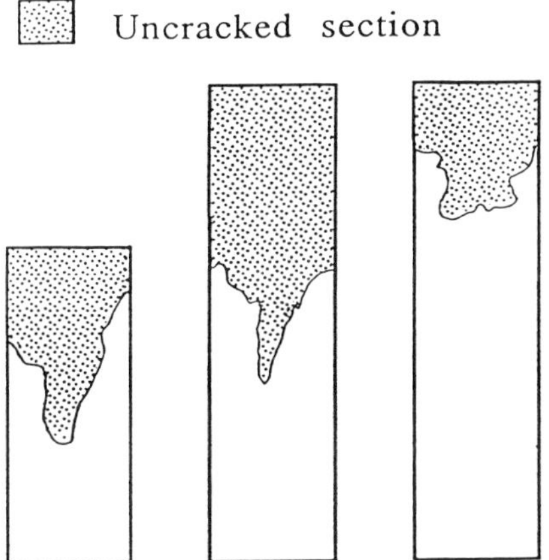

Fig. 2. Crack profiles revealed from dye-penetration technique (from [33]).

of crack front is believed to be due to the heterogeneity of concrete and possible three-dimensional stress states along the crack front. As a result, care should be exercised when correlations between model predicted surface crack lengths and experimental measurements are attempted.

e. Strain softening

Unlike metal which has strain hardening behavior, concrete is reported to have softening behavior as indicated in Fig. 3, in which the stress-displacement curves of several uniaxial direct tensile tests are plotted [14]. The softening behavior occurred after a non-linear ascending stress-strain relationship. After extensive study by several researchers, it was found that strain measurements after the peak stress do not have a unique value and are dependent on the gauge length. Nevertheless, a post-peak stress-separation relationship was found to be a better representation of the post-peak softening behavior of concrete. The uniqueness of this stress-separation relationship, however, is still an open question and more research is needed to resolve this problem.

The above observations described some unique features observed on the quasi-brittle fracture behavior of concrete. Due to the complexities involved, it is a difficult task to model all the above mentioned mechanisms in one model. Therefore, depending on the variables that are of interest to engineers, different models have been proposed. In the following sections, several fracture models are discussed and validity criteria for evaluation of fracture models are presented.

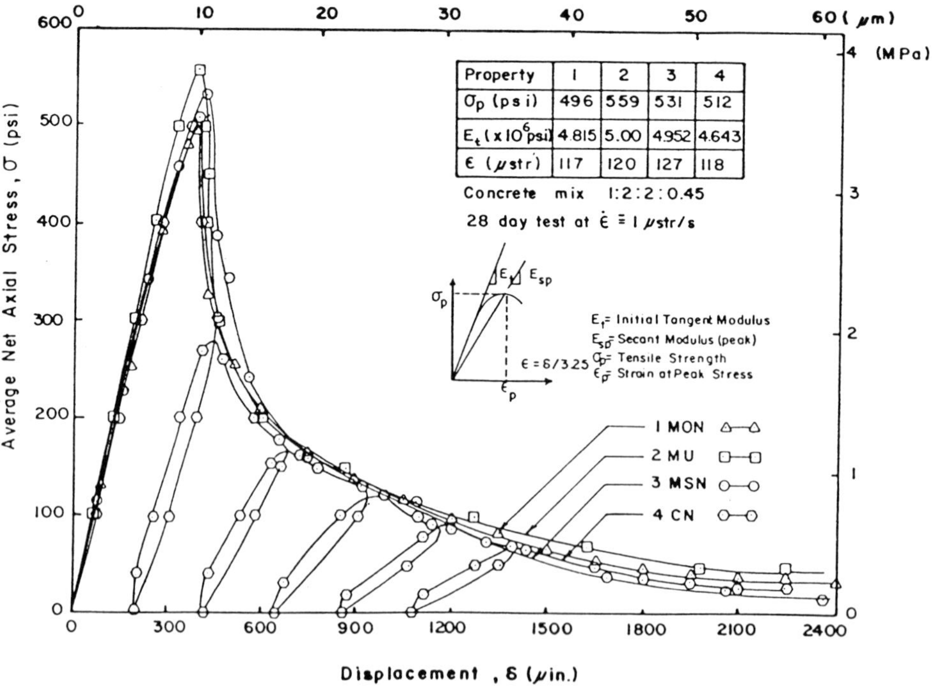

Fig. 3. Typical stress-displacement results of uniaxial tensile tests (from [13]).

3. Mathematical modeling

In the present paper, only models containing failures caused by crack propagation will be discussed. A crack subjected to applied mixed-mode loads with bridging stress acting on part of the crack surface is presented in Fig. 4. For simplicity only the planar case is discussed here. Assuming that materials outside the process zone are linear elastic, the energy release rate (G) (or specific energy dissipation) during crack propagation can be expressed as

$$G = \frac{K_I^2}{E'} + \frac{K_{II}^2}{E'} + \int_0^{CTOD} \sigma \, dw + \int_0^{CTSD} \tau \, ds + \int_\Gamma W \, dy, \tag{1}$$

in which K_I and K_{II} are the net stress intensity factors for modes I and II; σ and τ are respectively the normal and shear traction acting on the crack surfaces; these traction forces are assumed to be functions of crack tip opening displacement (w) and crack tip sliding displacement (s); $E' = E$ for plane stress case; $E' = E/(1 - v^2)$ for plane strain case; $E = $ Young's modulus; $v = $ Poisson's ratio; Γ is the integration contour along the microcrack band; and W is the strain energy density with strain softening behavior along a band of microcracks. Equation (1) can be derived from the J-integral [33] and other methods. It can be seen that the Griffith–Irwin energy dissipation mechanism is represented by non-zero stress intensity factors and the Dugdale-Barenblatt energy dissipation mechanism is expressed by the two traction terms. The physical interpretation of Dugdale–Barenblatt energy dissipation originated from atomic force consideration (Barenblatt) and yielding of materials (Dugdale). Existence of a band of softening (or damage) zone is based on the formation of a microcrack band along crack surfaces during crack propagation.

Although energy dissipation during crack propagation is contributed by different mechanisms, there is no acceptable method to determine the weight of each mechanism. However, (1) does provide a mathematical foundation for formulation of different energy-based models, i.e., one can formulate different fracture mechanics models using the associated variables, K_I, K_{II}, σ, τ, and W.

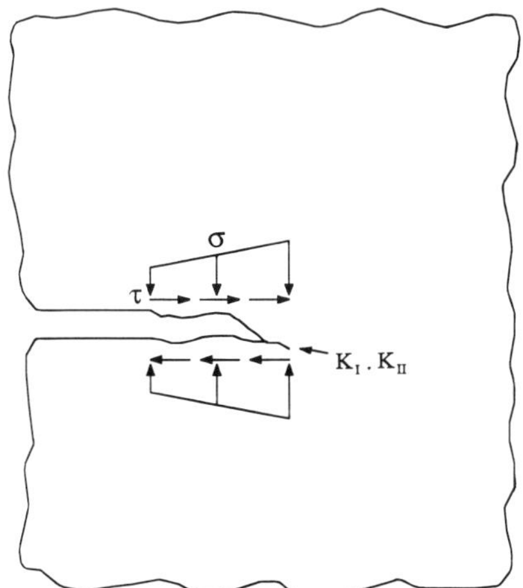

Fig. 4. A crack subjected to mixed-mode loading.

4. Mode I fracture models

To study mode I fracture, a hybrid model was proposed by Jenq and Shah [20]. A closing pressure, which results from the physical observations of aggregate-interlock along crack surfaces, and a non-zero stress intensity factor at the crack tip were assumed. The energy release rate associated with these two mechanisms can be derived from (1) by eliminating all the terms associated with shear (K_{II} and τ) and the microcrack band.

$$G = \frac{K_I^2}{E'} + \int_0^{CTOD} \sigma \, dw. \tag{2}$$

This model can be used to explain the reported toughening mechanism and energy dissipation provided by aggregate-interlock. However, the weight of energy contribution due to aggregate-interlock is difficult to assess [20].

Although the hybrid model presents a more general approach, it is not as computationally efficient as compared to models that assume a single energy dissipation mechanism, i.e. effective Griffith crack models or the Dugdale-Barenblatt type of cohesive crack models.

For example, following the Dugdale-Barenblatt concept, Hillerborg et al. [15] proposed a cohesive crack model (or Fictitious Crack Model) to simulate crack propagation in concrete. In this model, the stress intensity factor is assumed to be zero and all the fracture energy is assumed to dissipate through a strip of fictitious crack. In the model, a general stress-strain relationship is proposed for the undamaged material; however, a linear stress-strain relationship (Fig. 5a) is always assumed for computational convenience. A crack is assumed to initiate and propagate when the principal tensile stress reaches the tensile strength. Dissipation of mode I fracture energy is achieved through a unique stress-separation curve, which can be obtained from a uniaxial tensile test (Fig. 3) and further idealized as indicated in Fig. 5. Based on (1), the specific energy dissipation up to a certain opening can be expressed as

$$G = \int_0^{CTOD} \sigma \, dw, \quad \text{for the cohesive crack model.} \tag{3}$$

It can be further demonstrated that fracture energy, which is defined as the energy required to

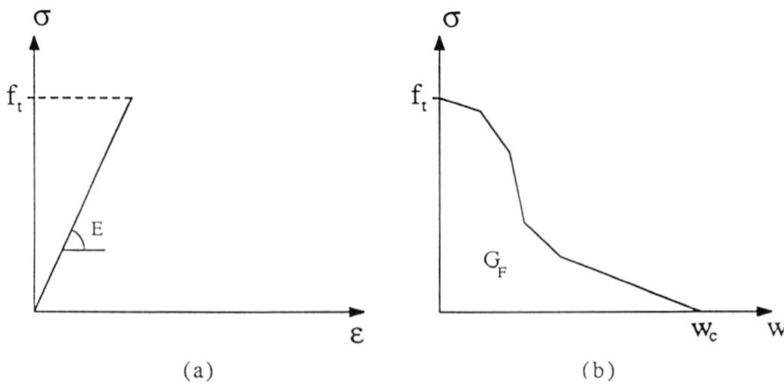

Fig. 5. Fictitious crack model.

produce a unit area of traction free crack can be expressed as

$$G_F = \int_0^{w_c} \sigma \, dw \tag{4}$$

in which w_c is the critial crack tip opening displacement beyond which stress transfer capability is negligible. If the stress-separation curve (Fig. 5b) is assumed to be unique, then G_F is also a material constant. Based on the stress-separation relationship, crack resistance of a concrete structure can be determined from global force equilibrium. It should be noted that three fracture properties, tensile strength, fracture energy, and shape of the stress-separation curve, are needed for the Fictitious Crack Model. Up to the present time, there is not a unique method that can be equivocally accepted to determine these three fracture properties; the uniqueness of the material properties associated with this model is difficult to verify [23]. One advantage of this model however is that a complete load-displacement behavior can be simulated.

By dissipating the fracture energy through a strain softening relationship and a band of damage (Fig. 6), a Crack Band Model was proposed by Bazant and Oh [3]. Stress continuity is assumed, there is no stress singularity, and the energy dissipation during crack growth can be calculated as

$$G = \int_\Gamma (W \, dy), \quad \text{for the crack band model.} \tag{5}$$

It should be noted that the crack band should be completely included in the integration contour, Γ, in (5). To maintain a constant fracture energy, the strain softening relationship and the band width of the damage zone are coupled and have to be related to the fracture energy. Again, global force equilibrium can be used to calculate the crack resistance during progressive development of damage.

By modifying the Griffith-Irwin energy release rate concept, a Two Parameter Fracture Model (TPFM) was proposed by Jenq and Shah [20]. Due to the existence of air voids, initial flaws, weak interfaces, and tougher aggregates in concrete, there is always pre-critical stable crack growth before the peak load is reached. This pre-critical stable crack growth is accompanied by

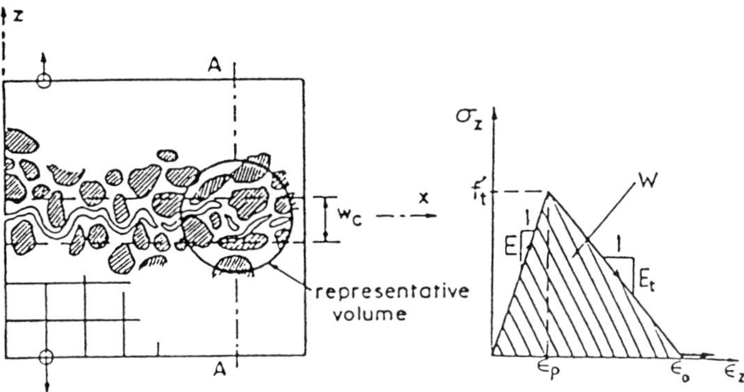

Fig. 6. Crack band model.

R-curve behavior during progressive crack development [40]. An effective traction-free crack is assumed in this model. For mode I fracture the energy release rate can be expressed as

$$G = \frac{K_I^2}{E'}, \quad \text{for the effective Griffith crack model.} \tag{6}$$

It was proposed that a crack will propagate at a critical stress intensity factor (K_{Ic}^s) when the critical load is reached. In addition, at the critical load, the crack tip opening displacement, which is measured at the original crack tip, is assumed to be a material constant. Therefore, the critical energy release rate (G_c) after the critical load is reached can be expressed as

$$G_c = (K_{Ic}^s)^2/E'. \tag{7}$$

A three-point bend notched beam test [28] can be used to determine these two fracture parameters. Once these two parameters are known, crack resistance of a concrete structure can be calculated [22] by satisfying the following conditions: the stress intensity factor at the effective crack tip is equal to K_{Ic}^s and the crack tip opening displacement at the original crack tip is equal to $CTOD_c$ (Fig. 7).

The above discussion summarizes several energy-based fracture models. It should be noted that strength-based models, e.g. plasticity-based models, and continuum damage models, which are not discussed in the present paper, have also been proposed to characterize the fracture behavior of concrete. Results predicted from these models are, of course, different. However, there is no method that can be used to objectively evaluate the validity of different theoretical models. In the following section, a set of criteria for evaluating concrete fracture models is discussed and the Two Parameter Fracture Model is screened against these criteria.

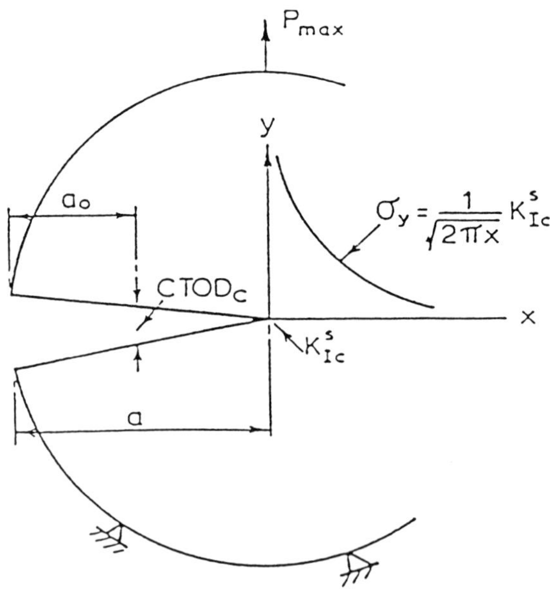

Fig. 7. Two Parameter Fracture Model.

5. Validation criteria for concrete fracture models

Since existence of microcracks, the three-dimensional crack profile, uniqueness of softening curves, surface roughness, and inclusion toughening mechanisms are difficult to quantify, these factors can not be directly used to evaluate the validity of fracture models. However, the above mechanisms are reflected either directly or indirectly in the experimentally observed load bearing capacity of concrete structures. Notch-sensitivity and effects of specimen size on the conventional critical stress intensity factor and on flexural strength of concrete are the frequently reported experimental trends. They can be used to evaluate the validity of fracture models. In addition, a valid fracture model should also satisfy 'portability' condition, i.e. the model should be applicable to a structure of an arbitrary geometry. Nevertheless, the associated fracture parameters should be determined in a self-consistent and unambiguous manner. Therefore, the following criteria are proposed for evaluating validity:

(a) notch sensitivity;
(b) size effect on conventional critical stress intensity factor;
(c) size effect on flexural strength;
(d) portability;
(e) determination of fracture parameters.

The above criteria will be discussed in the following sections and theoretical results of the Two Parameter Fracture Model are screened with these criteria.

Material length (Q) and theoretical tensile strength (f_t)

To normalize the theoretical predictions of the Two Parameter Fracture Model, material length (Q) and theoretical uniaxial tensile strength (f_t) of concrete are defined as

$$Q = (E'^* CTOD_c / K_{Ic}^s)^2 \qquad (8)$$

and

$$f_t = 1.4705(K_{Ic}^s)^2 / (E'^* CTOD_c). \qquad (9)$$

The material length (Q) represents a length measurement from a material's viewpoint. Therefore effects of specimen size or brittleness of a structure can be objectively presented when the absolute length is expressed in terms of the material length. Ranges for material length (Q) are 1.25 cm–5 cm (0.5″–2″), 5 cm–15 cm (2″–6″), and 15 cm–35 cm (6″–14″) for hardened cement paste, mortar, and concrete respectively.

Conventional critical stress intensity factor

It was reported by Higgins and Bailey [16] and other researchers that the conventional stress intensity factor ($\bar{K}_{Ic}$), which is calculated based on the measured peak load and the initial notch length, is dependent on the beam depth. When the depth of the specimen is large enough, $\bar{K}_{Ic}$ will approach a plateau value as shown in Fig. 8a. This demonstrates the invalidity of a single parameter LEFM approach. This size effect, however, is correctly predicted by the Two Parameter Fracture Model as shown in Fig. 8b.

Notch sensitivity

By measuring the net failure strength of notched beams and unnotched beams, Shah and McGarry [34] reported different degrees of notch-sensitivity for cement paste, mortar, and concrete. Concrete was found to be the least notch-sensitive material compared to mortar and hardened cement paste (Fig. 9a) for beams of the same size. It is clear that conventional strength-based models can not be used to model this effect.

A normalized theoretical prediction of notch sensitivity of concrete is given in Fig. 9b in which the specimen size is represented in terms of material length (Q). It can be seen that materials with smaller material length are notch sensitive and more brittle if the beam depth is kept constant. Since concrete has a larger value of material length (15 cm to 35 cm), concrete will be less notch sensitive compared to hardened cement paste and mortar which have much smaller material lengths as reported earlier.

Size effect on flexural strength

It has long been recognized that flexural strength (or modulus of rupture) of a concrete beam decreases as the beam depth increases; this flexural strength will approach a constant value if the specimen depth is large enough. This size effect was empirically correlated with beam depth by CEB-FIB code based on numerous experimental results as indicated in Fig. 10. Again, conventional strength-based models or a single parameter LEFM model can not predict this size effect.

By assuming that the material length is equal to 20 cm (8 inches), which is the average material length of concrete and mortar [18], the effect of beam depth on flexural strength was predicted and is plotted in Fig. 10. The CEB-FIP size effect compares favorably with the theoretical prediction. This also demonstrates the model's capability in predicting the fracture resistance of notched as well as unnotched structures.

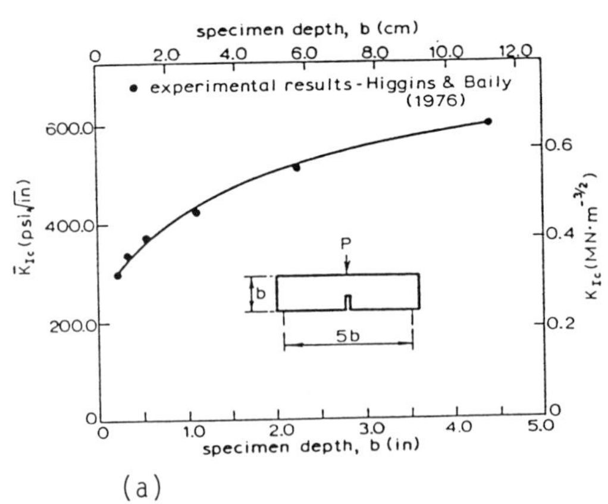

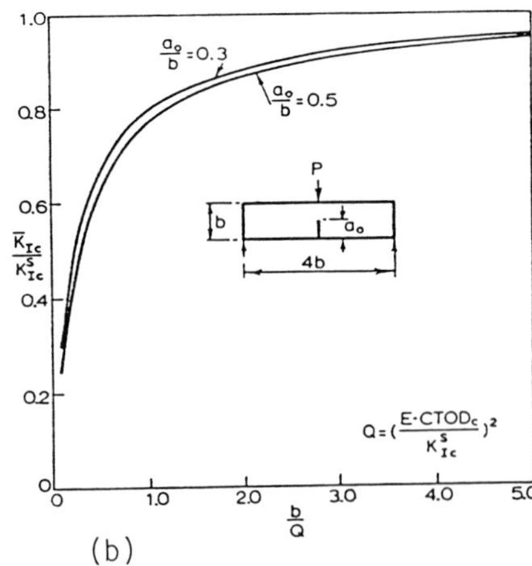

Fig. 8. Experimental results (a) and theoretical prediction (b) of effects of specimen size on conventional critical stress intensity factor.

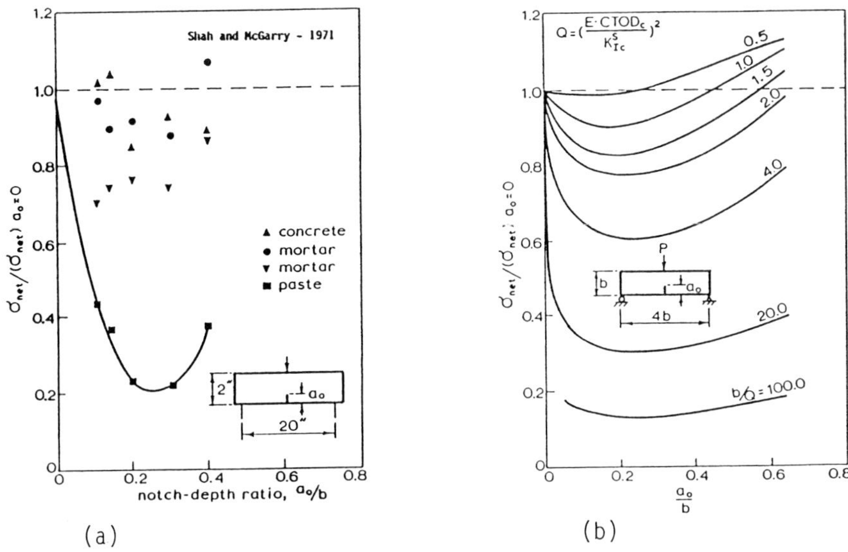

Fig. 9. Experimental results (a) and theoretical prediction of notch-sensitivity of concrete.

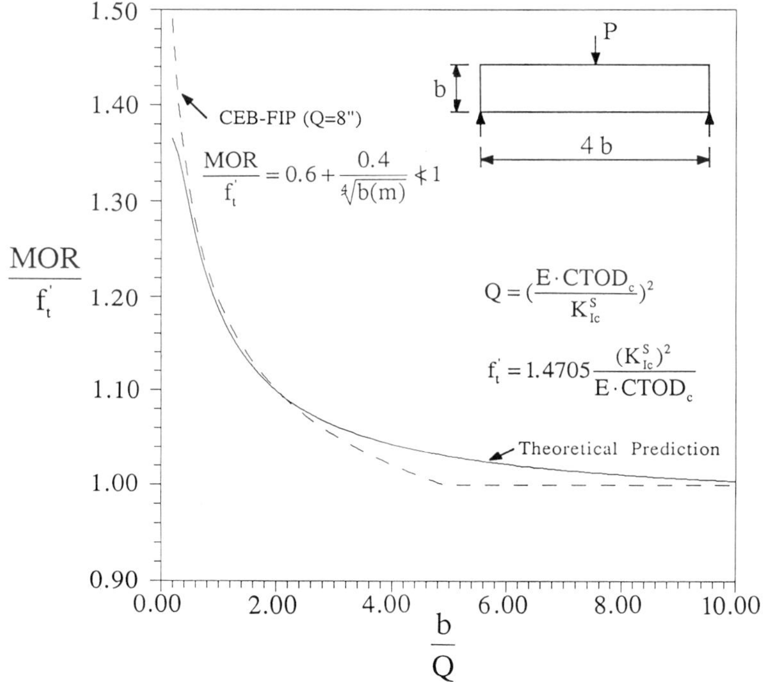

Fig. 10. Effect of specimen depth on modulus of rupture.

Portability and determination of fracture parameters

The Two Parameter Fracture Model was developed for structures of arbitrary shape [22] and has been applied to structures of different geometries, e.g. compact tension specimens, wedge splitting specimens, etc., [23]. Thus the portability condition of this model is satisfactorily met.

Young's modulus and fracture parameters associated with the Two-Paramater Fracture Model can be objectively determined from a three-point bend notched beam test [28]. Furthermore, since the method used in determining the material parameters is also consistent with the way these parameters are applied, self-consistency on the application of these parameters is also maintained.

The above discussion demonstrates validity evaluation on the Two Parameter Fracture Model. It can be further demonstrated that all the energy-based models, i.e. the Fictitious Crack Model, the Crack Band Model, and the Two Parameter Fracture Model, will meet the proposed validation criteria. It was shown by Planas and Elices [32] that the theoretical peak loads obtained from the above models are about the same for reasonable structure sizes, i.e. a size large enough that the homogeneity assumption still holds, if the associated fracture parameters are properly selected. This is not surprising since either a constant energy release rate (4) or a constant fracture energy (7) is always prescribed in the above models for a reasonably large sized structure.

The theoretical predictions of the above models are, however, very different for small sized structures. This can be due to the fact that for small sized structures, the above models may no longer be valid since the homogeneity assumption is no longer satisfied.

6. Mixed-mode fracture

Based on the same principle, the above mentioned cohesive crack models and effective Griffith crack models have been extended to study crack propagation under mixed-mode loading conditions [1, 2, 4, 5, 7, 8, 19, 21, 24, 29, 38, 39, 41]. For most of the models, only energy dissipation associated with tensile fracture was considered. This may be acceptable if it can be assumed that the contribution of shear terms is negligible.

To predict crack propagation path under mixed mode loading conditions, it is reasonable to assume that a crack will propagate along the direction that offers least resistance to the applied load. Fracture resistance can be determined using different variables, e.g. tensile stress, energy release rate, or strain energy density. Based on this assumption, several theories have been proposed to determine the trajectory of a propagating crack. Maximum tensile stress criterion, maximum energy release rate criterion, and minimum strain energy density criterion are widely used [24]. Figure 11 shows the experimentally observed crack paths of four-point shear tests (Fig. 12) [6, 8, 24] along with the theoretical predictions. Crack paths predicted using different criteria were found to be almost identical and in good agreement with experimental results. Therefore, it can be concluded that crack paths can be satisfactorily predicted using the above mentioned criteria.

To determine the crack resistance at different failure paths under mixed-mode loading, the Two Parameter Fracture Model was extended by replacing the mode I stress intensity factor and crack tip opening displacement with mixed-mode stress intensity factors and crack tip displacements, Fig. 13 [21].

Four-point shear tests were analyzed using finite elements methods. Two possible failure paths (one at the location of notch and the other at the location of maximum moment, i.e., path z in Fig. 12) were considered for $c/b = 0.8$ and $c/b = 1.2$. For $c/b = 1.2$ and $a/b = 0.2$, it was

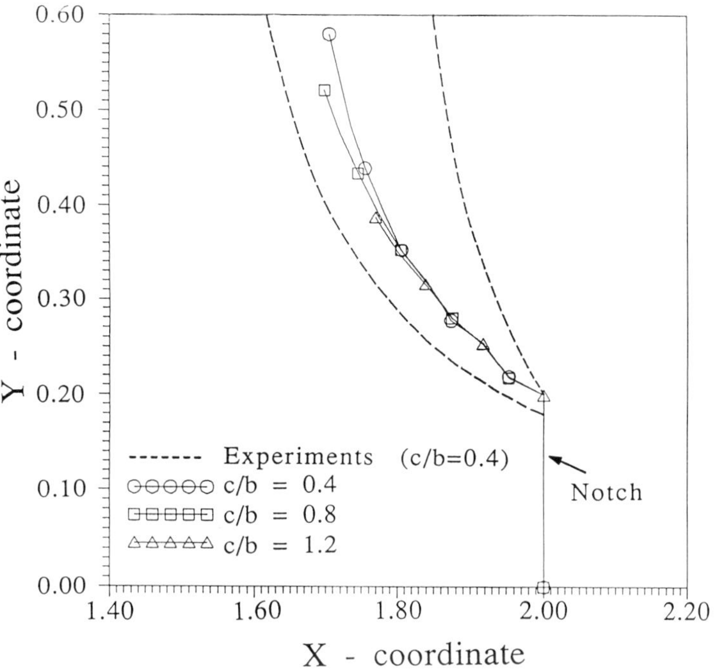

Fig. 11. Crack trajectories of four-point shear tests.

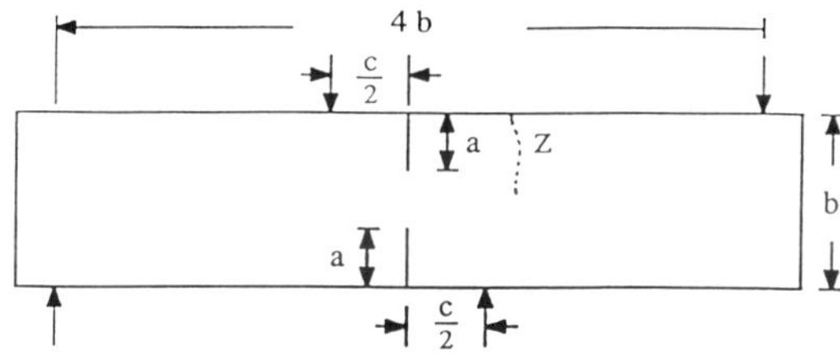

Fig. 12. Dimensions of the four-point shear tests.

found that the fracture resistance offered at the maximum moment location is much lower than that offered at the location of the notch by 50 percent to 80 percent. Therefore, the location at the maximum moment will dominate the final failure path for $c/b = 1.2$. This prediction was also observed in the experiments [7]. For $c/b = 0.8$, the load resistance offered by the two possible failure paths are within 10 percent of each other. With the heterogeneity of concrete materials, it is possible that final failure will occur at either location. It was, however, observed by Ballatore et al. [8] that most of the failures occurred at the location of the imposed notch except for one specimen. Therefore, it is very clear that $c/b = 0.8$ is close to the transition point at which the final failure location may shift from one location to the other. Figure 14 shows the model prediction of the effect of specimen size on the peak stress values. Similar mixed-mode size-effect was also reported by Ballatore et al. [8] in their experiments. Good agreement was also found between the theoretical prediction and experimental results of peak load values on the four-point shear tests (Fig. 15).

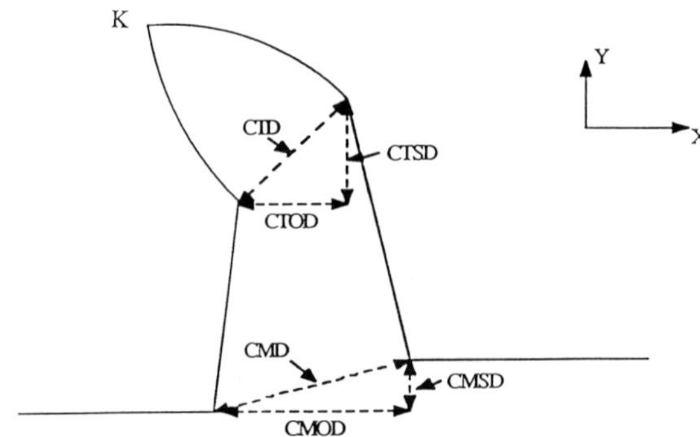

Fig. 13. Extension of Two Parameter Fracture Model to mixed-mode case.

With a unique set-up, Xu and Reinhardt [41] have performed torsional tests on notched concrete cylinders to investigate energy consumed when a crack is propagating under an applied torsion. It was reported that the fracture energy is a function of the specimen size and notch-depth ratio. They attribute this geometrical effect to possible frictional energy consumed along crack interfaces. Since frictional energy consumed during the fracturing process is dominated by specimen geometry and loading paths and is not a material constant, the shear term (τ) in (1) can not be simulated using a unique stress-separation curve as proposed in the mode I cohesive crack model (Fig. 5). Instead, a Mohr-Coulomb type of frictional law should be applied. Similar results were also reported by Divarkar, Fafitis, and Shah [11], and van Meir [39]. This result indicated that possible frictional forces and the associated energy dissipation

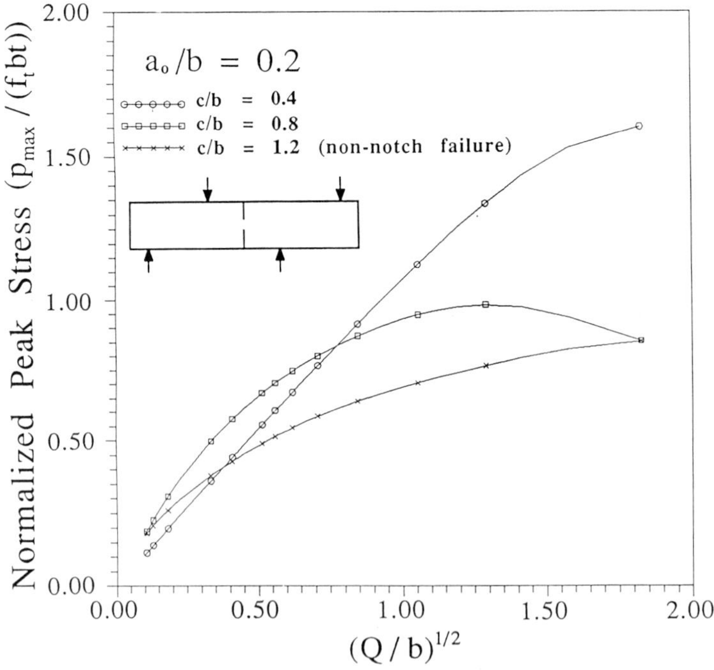

Fig. 14. Theoretical prediction of effect of size on four-point shear tests.

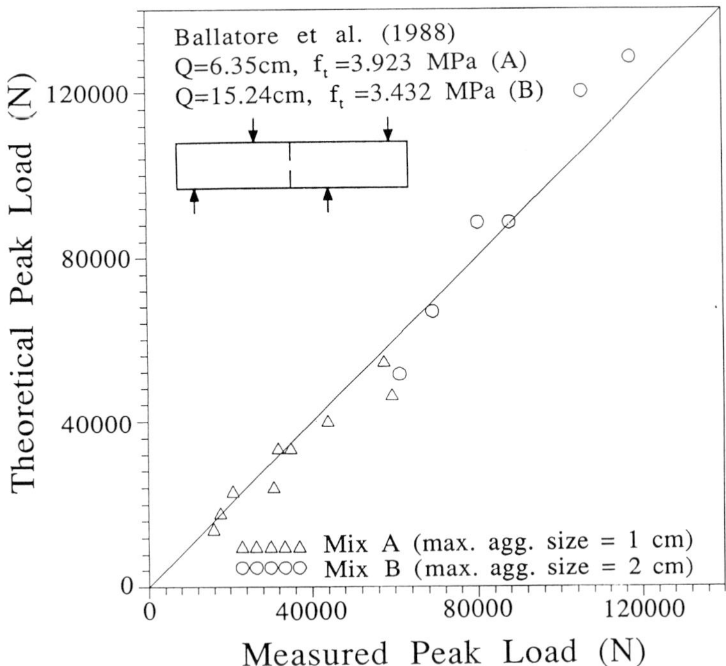

Fig. 15. Comparisons between the theoretical prediction and experimental results of four-point shear tests.

generated from the applied confining pressure have to be considered during mixed-mode crack propagation. Bazant, Prat, and Tabbara [6] have also reported similar results. However, they attribute the observed size effect to mode III fracture. More experimental data and theoretical development are certainly needed in order to identify the governing constitutive laws on the frictional force of concrete along the partially fractured crack surface.

7. Effects of loading rate

Since formation and propagation of micro-cracks and macro-cracks are a time dependent process, fracture behavior of concrete was found to be dependent on the loading rate [5, 13, 25, 26]. Available experimental data obtained by John and Shah [26] shows that critical stress intensity factor and Young's modulus are not very sensitive to the applied loading rate (see Figs. 16a and 16b) and can be reasonably assumed to be constant for loading rates between 10^{-7}/sec and 1/sec. The critical crack tip opening displacement, on the other hand, was found to be more sensitive to the loading rate (Fig. 16c), and decreases as the loading rate increases. This observation is in agreement with the widely accepted concept proposed in dynamic fracture and was later confirmed by FEM analysis [13]. Based on the available rate-dependent tensile strength data reported by different researchers, an empirical relationship was established to link the loading rate and the critical crack tip opening displacement [25]. Knowing the rate-dependent critical crack opening displacement and assuming that the critical stress intensity factor and Young's modulus are independent of the strain-rate, one can predict the effects of loading rate on fracture resistance of concrete structures. For example, Fig. 17 shows the theoretical prediction of the failure load as well as the failure mode of three-point bend

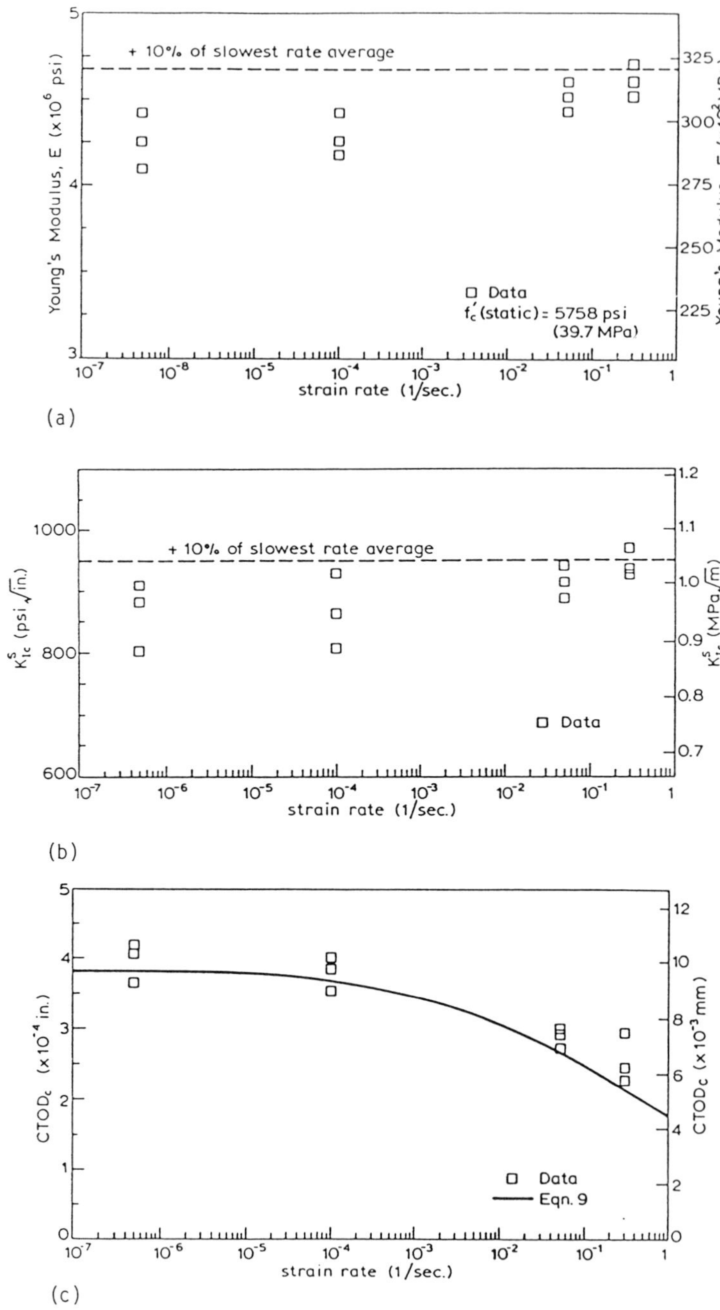

Fig. 16. Effects of loading rate on Young's modulus, critical stress intensity factor, and critical crack tip opening displacement.

mixed-mode tests. As the loading rate increases, the failure load also increases. Furthermore, there is a small window of off-set ratio (γ) that suggests the final failure mode will shift from notch to midspan for the same specimen geometry at different loading rates. These effects were clearly predicted by the rate-dependent model as given in Fig. 17. The theoretical prediction was

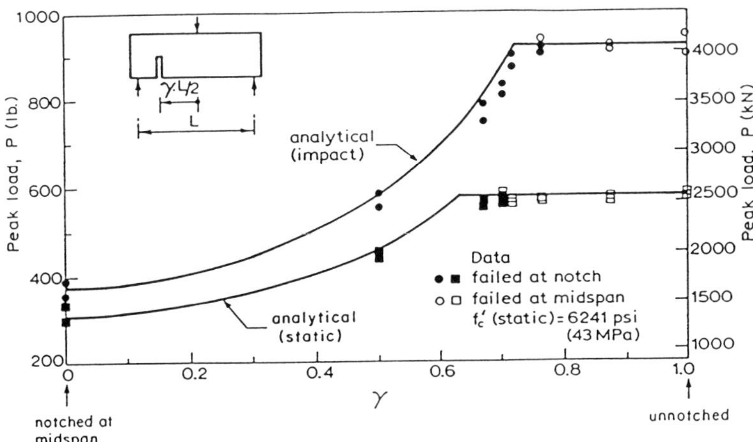

Fig. 17. Effect of strain rate on the failure load and failure mode of three-point bend mixed-mode tests.

found to be in good agreement with the experimental results obtained at different loading rates (Fig. 17) [26].

8. Conclusions

1. Features of quasi-brittle crack propagation are summarized in the present paper. Formulations of different energy-based fracture models were discussed based on their energy dissipation mechanisms. Notch sensitivity, size effect, and portability are proposed as validity criteria for evaluation of different concrete fracture models.
2. All the energy-based fracture models discussed in the present paper can be shown to satisfy the proposed validity criteria. Furthermore, these models can be traced back to the original constant Griffith surface energy concept.
3. Effects of mixed-mode loading conditions and effects of loading rates can be reasonably predicted from the extended Two Parameter Fracture Model. However, effects of confining pressure on mixed-mode crack propagation is still not very clear and more research is needed in this area.

Acknowledgements

Support provided by the National Science Foundation to the Center for Science and Technology of Advanced Cement-Based Materials (ACBM) is gratefully appreciated. Additional support provided by the Air Force Office of Scientific Research is also acknowledged. The first author appreciates the support provided by the Department of Civil Engineering at The Ohio State University in preparation of this manuscript.

References

1. M. Arrea and A.R. Ingraffea, Mixed-mode Crack Propagation in Mortar and Concrete, Report No. 81–13, Department of Structural Engineering, Cornell University, Ithaca, New York.
2. B.I.G. Barr and M. Derradj, in *Proceedings of International Conference on Fracture and Damage of Concrete and Rock*, Vienna, July 4–6, 1988.

3. Z.P. Bazant and B.H. Oh, *Materiaux et Constructions* 16, 93 (1983) 155–177.
4. Z.P. Bazant and P.A. Pfeiffer, *Materials and Structures*, RILEM 19 (1986) 111–121.
5. Z.P. Bazant and R. Gettu, in *Proceedings of International Conference on the Recent Developments of Fracture of Concrete and Rock*, Cardiff, UK, Sept. 1989, 549–565.
6. Z.P. Bazant, P.C. Prat and M.R. Tabbara, *ACI Materials Journal* 87, No. 1 (1990) 12–19.
7. A. Bocca, A. Carpinteri and S. Valente, in *Proceedings of International Conference on Fracture and Damage of Concrete and Rock*, Vienna, July 4–6, 1988.
8. E. Ballatore, A. Carpinteri, G. Ferrara and G. Melchiarri, *ibid.*
9. A. Carpinteri, *Engineering Fracture Mechanics* 16 (1982) 467–481.
10. A. Carpinteri and A.R. Ingraffea (eds), *Fracture Mechanics of Concrete: Material Characterization and Testing*, Martinus Nijhoff Publishers (1984).
11. M.P. Divarkar, A. Fafitis and S.P. Shah, *Journal of Structural Engineering* ASCE 113 (1987) 5.
12. K.T. Faber and A.G. Evans, *Acta Metallurgica* 31, No. 4 (1983) 565–576.
13. J. Du, A.S. Kobayashi and N.M. Hawkins, *Journal of Engineering Mechanics* ASCE 115, 10 (1989) 2136–2149.
14. V.S. Gopalaratanam and S.P. Shah, *ACI Journal* 82 (1985) 310–323.
15. A. Hillerborg, M. Modeer and P.E. Petersson, *Cement and Concrete Research* 6 (1976) 773–782.
16. D.D. Higgins and J.E. Bailey, *Journal of Material Science* 11 (1976) 1995–2003.
17. X.Z. Hu and F.H. Wittman, in *Proceedings of International Conference on the Recent Developments in Fracture of Concrete and Rock*, Cardiff, UK (1989) 307–316.
18. J.W. Hutchinson, *Journal of Applied Mechanics* 50 (1983) 1042–1051.
19. A.R. Ingraffea and W.H. Gerstle, in *Application of Fracture Mechanics to Cementitious Composites*, S.P. Shah (ed.), Kluwer Academic Publishers, Dordrecht, The Netherlands (1985).
20. Y.S. Jenq and S.P. Shah, *Engineering Fracture Mechanics* 21, No. 5 (1985) 1055–1069.
21. Y.S. Jenq and S.P. Shah, *International Journal of Fracture* 38 (1988) 123–142.
22. Y.S. Jenq and S.P. Shah, *Journal of Engineering Mechanics*, ASCE 111, 4 (1985) 1227–1241.
23. Y.S. Jenq and S.P. Shah, Geometrical Effects on Mode I Fracture Parameters. Report to RILEM Committee 89-FMT, January 1989.
24. Y.S. Jenq and S.P. Shah, in *Proceedings of International Conference on the Recent Developments of Fracture of Concrete and Rock*, Cardiff, UK (1989) 27–38.
25. R. John, S.P. Shah and Y.S. Jenq, *Cement and Concrete Research* 17 (1987) 249–262.
26. R. John and S.P. Shah, *Journal of Structural Division*, ASCE, to appear.
27. B.L. Karihaloo and P. Nallathambi, *Cement and Concrete Research*, to appear.
28. B.L. Karihaloo and P. Nallathambi, Final Report, A. Notched Beam Test: Mode I Fracture Toughness, Sub-committee of RILEM TC89-FMT, April, 1988.
29. A.S. Kobayashi, N.M. Hawkins, D.B. Barker and B.M. Liaw, in *Application of Fracture Mechanics to Cementitious Composites*, S.P. Shah (ed.), Martinus Nijhoff Publishers (1985) 25–50.
30. A. Maji and S.P. Shah, *Experimental Mechanics* (1988) 27–33.
31. R. Miller, S.P. Shah and H.I. Bjelkhagen, *Experimental Mechanics* (1988) 388–394.
32. J. Planas and M. Elices, in *Proceedings, CRS-NSF International Workshop on Size Effect and Strain Localization due to Cracking and Damage*, Cachan, France, Sept. 1988.
33. J.R. Rice, *Journal of Applied Mechanics* 35 (1968) 379–386.
34. S.P. Shah and F.J. McGarry, *Journal of Engineering Mechanics Division*, ASCE (1971) 1663–1675.
35. S.P. Shah and F.L. Slate, in *International Conference on the Structure of Concrete and Its Behavior under Load*, London, 1965, Cement and Concrete Association.
36. S.E. Swartz and T. Refai, in *Proceedings of SEM/RILEM International Conference on Fracture of Concrete and Rock*, Houston, Texas (1987) 403–417.
37. S.E. Swartz and T. Refai, in *Proceedings of International Workshop on Fracture Toughness and Fracture Energy–Test Methods for Concrete and Rock*, Sendai, Japan (1988).
38. N. Taha and S. Swartz, in *Proceedings of International Conference on Recent Developments on the Fracture of Concrete and Rock*, Cardiff, U.K. (1989) 5–17.
39. J.G.M. van Mier, in *Proceedings of SEM Spring Conference on Experimental Mechanics*, Cambridge, Massachusetts (1989) 51–58.
40. M. Wecharatana and S.P. Shah, *Journal of Engineering Mechanics* (1982) 1100–1113.
41. D. Xu and H.W. Reinhardt, in *Proceedings of International Conference on the Recent Development of Concrete and Rock*, Cardiff, UK (1989) 39–50.

International Journal of Fracture **51**: 121–138, 1991.
Z.P. Bažant (ed.), Current Trends in Concrete Fracture Research.
© 1991 *Kluwer Academic Publishers. Printed in the Netherlands.*

Size dependence of concrete fracture energy determined by RILEM work-of-fracture method

ZDENĚK P. BAŽANT and MOHAMMAD T. KAZEMI
Center for Advanced Cement-Based Materials, Northwestern University, Evanston, Illinois. 60208, USA

Received 1 June 1990; accepted 1 November 1990

Abstract. The paper analyzes the size dependence of the fracture energy of concrete obtained according to the existing RILEM recommendation proposed by Hillerborg and based on the work-of-fracture method of Nakayama, Tattersal and Tappin, in which the energy dissipated at the fracture front is evaluated from the measured load-displacement curve. The analysis is based on the size effect law proposed by Bažant, which has been shown to be applicable to the size ranges up to about 1:20 and apply in the same form for all specimen geometries. The analysis utilizes the previously developed method for calculating the R-curve from the size effect, and the load-deflection curve from the R-curve. The R-curve is dependent on the geometry of the specimen. The results show that the fracture energy according to the existing RILEM recommendation is not size-independent, as desired, but depends strongly on the specimen size. This dependence is even stronger than that of the R-curve. When the specimen size is extrapolated to infinity, the fracture energy according to the RILEM recommendation coincides with the fracture energy obtained by the size effect method. It is also found that, in fracture specimens of usual sizes, the pre-peak contribution of the work of the load to the fracture energy is relatively small. Finally, as a by-product, the analysis also verifies the fact that, in three-point bend fracture specimens, the fracture energy according to the RILEM definition is dependent on the notch depth.

1. Introduction

Since concrete is a brittle heterogeneous material in which fracture is preceded by a large fracture process zone of variable size, determination of the fracture energy of concrete, G_f, is not an easy matter. The only unambiguous definition of fracture energy can be given in terms of an extrapolation of the specimen size to infinity [1–2]. This definition of course requires knowledge of the size effect law for fracture. Its exact form is not known, but an approximate size effect law for the nominal stress at failure that is sufficient for most practical purposes has recently been established [1–4]. When the existing RILEM recommendation for the measurement of fracture energy of concrete was formulated [5–6], the size effect law for fracture had not yet been known. The fracture energy was in that recommendation defined according to the work-of-fracture method, originally proposed by Nakayama [7], and by Tattersall and Tappin [8] (the latter authors also discovered size effect in this method). Adoption of the work-of-fracture method for concrete was recommended by Hillerborg and co-workers [6, 9–10] on the basis of their work as well as earlier studies of concrete (e.g. [10–12]).

Based on a measured load-deflection curve of a fracture specimen, typically a three-point-bend beam, the work of load P (including the effect of its own weight) on the load-point displacement u is, in the RILEM method, calculated as

$$W_f = \int_0^{u_1} P\,du, \tag{1}$$

in which u_1 = final displacement at which the load is reduced to zero. The fracture energy according to the RILEM definition, G_f^R, represents the average (or effective) fracture energy in the ligament [7, 11]. It is obtained as [5–6, 9–10]:

$$G_f^R = W_f/bl, \quad l = (1 - \alpha_0)d, \tag{2}$$

in which l = length of ligament, d = depth of the beam, $\alpha_0 = a_0/d$, a_0 = length of notch, b = thickness of the beam (Fig. 1), and bl = effective area of fractured ligament. However, as we will see, the averaged value depends considerably on the specimen size as well as shape. Such differences would have to be minor if the fracture process zone (or cohesive zone, bridging zone) were very small compared to the ligament length. However, this is not the case in fracture testing of concrete (except if impractically large specimens were used).

The RILEM definition is based on Hillerborg's fictitious crack model [6, 10, 13] and previous similar models for other materials (e.g. [14]) in which it has been tacitly assumed that the crack bridging stress σ_b in the cohesive zone (or fracture process zone) is a function of the crack opening displacement δ but no other variable, i.e. that the relation of σ_b and δ is unique. If this hidden hypothesis were true, then (2) would have to be exact, independent of specimen size, because the total energy dissipated along the ligament (i.e. $\int_0^\infty \sigma_b \, d\delta$) would be uniform (as the complete curve $\sigma_b(\delta)$, down to zero σ_b, has been traced at every point of the ligament by the time of complete fracture). As we will see, however, this hypothesis cannot be true. In fact there is no reason to expect the path $\sigma_b(\delta)$ not to depend on the normal strains parallel to the crack plane, on the deformation state in a small neighborhood of the point, and on the deformation history.

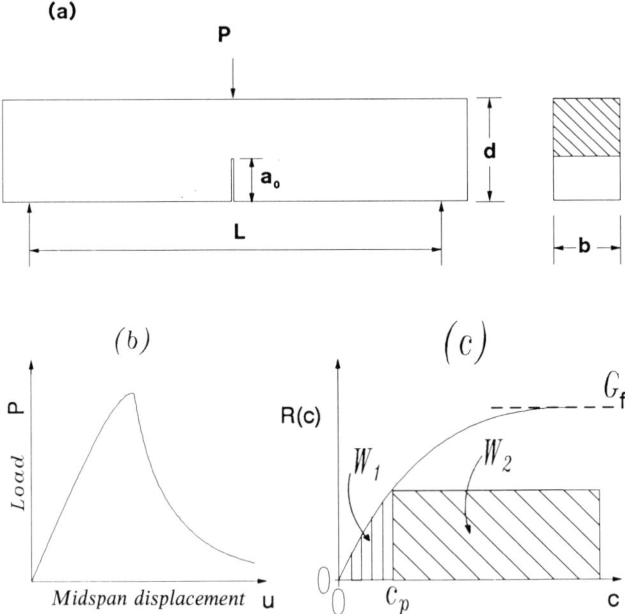

Fig. 1. Three-point bend specimen geometry, load deflection curve, R-curve and subdivision of work according to the R-curve.

Note that the fictitious crack model does not imply the energy release rate G for the elastically equivalent sharp crack to be constant during crack propagation. The reason is that, unless the fracture front is remote from both the notch tip and specimen boundaries, the σ_b-distribution throughout the cohesive zone varies during propagation, giving different values for G; $G = \int_0^l \sigma_b [\delta(x)][d\delta(x)/da]\,dx$ where x is the coordinate along the ligament measured from the notch tip, and $d\delta/da = (d\delta/dt)/(da/dt)$ where $t = $ time and $a = $ depth of effective (elastically equivalent) crack (which gives the same specimen compliance).

As a variant of the RILEM method, the fracture energy can also be measured by fracturing the specimen only partially. In this procedure, which usually involves unloading in the softening zone [15], the measured work (partial work-of-fracture) is divided by the area of the advancing crack, which needs to be somehow measured or estimated. As another variant, some researchers used cyclic testing for calculation of the fracture energy as a function of crack length [16–17]. In this method, the partial works-of-fracture between any two consecutive cycles are divided by the corresponding areas of crack advance. It needs to be noted, however, that the loading rate [17–18] and creep of course affect the work-of-fracture. This effect is greater in cyclic than monotonic tests. Moreover, cyclic loading no doubt produces inelastic volume dilatancy and residual stress in the fracture process zone, which are absent in monotonic tests.

Various series of experiments have demonstrated that G_f^R is significantly size and shape dependent [8–9, 17–32]. Planas, Elices, and co-workers [28–30] attempted to extrapolate the values of G_f^R to infinite size ($G_f^P = \lim G_f^R$ for $d \to \infty$), using a simple and approximate size effect relation, which they called the perturbed ligament model:

$$G_f^R = G_f^P \frac{l - l_p}{l}, \tag{3}$$

where $l_p = $ a size-independent constant called the perturbation length (which depends on material properties as well as geometry), and $l = $ ligament length, which appears to be a few-times larger than l_p. Equation (3) is treated as independent of specimen geometry, although this cannot be exactly so since otherwise two different specimens of the same l, e.g. three-point-bend and wedge-splitting specimens, would have to yield the same fracture energy. Parameters G_f^P and l_p can be obtained by linear regression analysis. Planas, Elices, and co-workers [28–30] concluded that the fracture energy values so obtained appear to be unambiguous, and that G_f^P needs to be determined from a set of tests of specimens of different sizes, with a sufficient size range, similar to the size effect method proposed by Bažant [1, 2].

The purpose of this study is to analyze the size dependence of G_f^R on the basis of the size effect law proposed by Bažant [3] for the nominal strength at failure. This has recently been shown to have a relatively broad validity, being applicable to the size range up to about $1:20$, and to be rather insensitive to the specimen shape (geometry).

2. Review of size effect law

The most important consequence of fracture mechanics is the effect of size on the nominal strength (i.e. nominal stress at maximum load). To describe it, we consider geometrically similar structures (or specimens) of different sizes (with geometrically similar notches or initial cracks), and introduce the nominal strength

$$\sigma_N = c_n \frac{P_u}{bd},\qquad(4)$$

where P_u = maximum load (ultimate load), b = thickness of two-dimensionally similar structures or specimens (which should best be chosen the same for test specimens of all the sizes), d = characteristic dimension of the structure or specimen, and c_n = coefficient introduced for convenience.

Plastic limit analysis, as well as elastic analysis with an allowable stress criterion or any failure criterion based on stress or strain, is known to exhibit no size effect, i.e., geometrically similar structures of different sizes fail at the same σ_N. However, this is not true in fracture mechanics. Due to similarity of stress fields in similar two-dimensional elastic structures of different sizes, the total potential energy of the structure must have the form $U = (\sigma^2/2E')Vf(\alpha)$ where $\sigma = c_n P/bd$, P = load; $V = bd^2$ = measure of volume; $f(\alpha)$ is a function of the relative crack length $\alpha = a/d$ (a is the crack length) and depends on the shape of the structure; $E' = E$ for plane stress, $E' = E/(1 - v^2)$ for plane strain, E = Young's modulus of elasticity, and v = Poisson's ratio. Therefore, the energy release rate is $G = -(\partial U/\partial a)/b = -(\partial U/\partial \alpha)/bd = -(\sigma^2/2E')f'(\alpha)d$, from which

$$G = \frac{P^2 g(\alpha)}{E'b^2d},\quad K_1 = \sqrt{GE'} = \frac{Pk(\alpha)}{b\sqrt{d}},\qquad(5)$$

where $f'(\alpha) = \partial f(\alpha)/\partial \alpha$, $g(\alpha) = -f'(\alpha)c_n^2/2$, and $k(\alpha) = \sqrt{g(\alpha)}$. The values of $g(\alpha)$ can be obtained by linear elastic analysis, such as elastic finite element analysis. For basic specimen geometries, $k(\alpha)$ also follows from the formulas for the stress intensity factors found in handbooks (e.g. Tada et al. [33]).

In materials with toughening mechanisms, such as rock, concrete, ice and ceramics, there is a large fracture process zone in front of the continuous crack. The maximum loads for various sizes may be assumed to occur when the tip of the equivalent elastic crack is at a certain distance c ahead of the tip of the initial crack or notch, i.e., $a = a_0 + c$ or $\alpha = \alpha_0 + c/d$ where $\alpha_0 = a_0/d$, and a_0 = initial crack or notch length. We assume that G cannot decrease and consider G_f to be the maximum value of G as $d \to \infty$. The corresponding maximum possible value of c at peak load, obtained for $d \to \infty$ (and $G \to G_f$), is denoted as c_f.

The fracture process zone starts with zero size and then grows as the loading increases while remaining attached to the notch tip (provided that the specimen geometry is such that $g'(\alpha) > 0$, which is the case for the three-point-bend specimen). The value of G required for fracture growth is basically determined by the size of the process zone. Since the value of c is also determined by this size, the value of G is directly related to the corresponding value of c. The magnitude of c at $P = P_u$ fully determines the value of α and, consequently, the value of $g(\alpha)$. So the ratio $G/g(\alpha)$ at $P = P_u$ for any specimen size should be approximately equal to $G_f/g(\alpha_1)$ where α_1 corresponds to c_f in a specimen of any size d, i.e. $\alpha_1 = (a_0 + c_f)/d = \alpha_0 + c_f/d$. Therefore, at $P = P_u$, we have $G = G_f g(\alpha)/g(\alpha_1)$. We may now substitute this expression into (5) along with $P_u \simeq \sigma_N bd/c_n$ and $g(\alpha_1) \simeq g(\alpha_0) + g'(\alpha_0)c_f/d$ (which ensues from Taylor series expansion assuming that $g'(\alpha_0) > 0$). Then, solving for σ_N, we obtain the size effect law [34]

$$\sigma_N = c_n \left(\frac{E'G_f}{g'(\alpha_0)c_f + g(\alpha_0)d} \right)^{1/2}. \tag{6}$$

Equation (6) may alternatively be written as [34–35]

$$\tau_N = \left(\frac{E'G_f}{c_f + \bar{d}} \right)^{1/2}, \tag{7}$$

where $\tau_N = \sqrt{g'(\alpha_0)}P_u/bd$, $\bar{d} = dg(\alpha_0)/g'(\alpha_0)$; τ_N = intrinsic nominal stress at failure, and $\bar{d}$ = intrinsic size of the structure, which are both independent of the specimen shape. The quantity that makes them shape-independent is the factor $g(\alpha_0)/g'(\alpha_0)$ [34], which has also been introduced for similar purposes by Planas and Elices [36] and for other purposes by Horii et al. [37].

Although the foregoing derivation [34] involves certain simplifications, other completely different arguments, including (1) dimensional analysis and similitude [3–4, 1], (2) simplified energy release considerations [3], and (3) deterministic limit of nonlocal generalization of Weibull-type statistical theory of strength [38], have been shown to yield the same result in a more general form:

$$\sigma_N = \frac{Bf_u}{\sqrt{1 + \beta}}, \quad \beta = \frac{d}{d_0}, \tag{8}$$

where f_u is a measure of material strength, B and d_0 are two empirical constants. This is equivalent to (6) if $B = c_n[E'G_f/c_f g'(\alpha_0)]/f_u$ and $d_0 = c_f g'(\alpha_0)/g(\alpha_0)$. Taking the limit of (5) in which $P = P_u = \sigma_N bd/c_n$, expressing σ_N from (8), and noting that $\lim \alpha \to \alpha_0$ for $d \to \infty$, one gets the formula [1]

$$G_f = \frac{B^2 f_u^2 d_0 g(\alpha_0)}{c_n^2 E'}. \tag{9}$$

Parameter B characterizes σ_N of very small structures and can be determined as $B = c_n P_u/f_u bd$ (from (8) for $\beta \to 0$), where P_u is calculated by plastic limit analysis. In the case that G_f is known the other parameter of the size effect law (d_0 in (8)) can be obtained from (9).

Parameter β in (8), called the brittleness number [1, 2], may be expressed as [34–35] either

$$\beta = \frac{B^2 f_u^2 g(\alpha_0)d}{c_n^2 E'G_f} \tag{10a}$$

or

$$\beta = \frac{g(\alpha_0)d}{g'(\alpha_0)c_f} = \frac{\bar{d}}{c_f}. \tag{10b}$$

Since (10a) involves the value of B that characterizes the peak load of a small structure, this equation is more accurate when β is small. On the other hand, (10b), based on (6), is more accurate when β is large since it is based solely on LEFM approximation.

Other definitions of the brittleness number were proposed by Homeny et al. [39], Carpinteri [40] and Hillerborg [6]. However, they only permit comparing specimens of different sizes but the same shape, and do not represent universal (shape-independent) measures of brittleness.

In an infinitely large specimen, the fracture process zone occupies only an infinitesimal volume fraction of the body, and so the body can be treated as elastic. Consequently, the stress and displacement fields surrounding the fracture process zone are the asymptotic elastic fields. They are known to be the same for any specimen geometry, and so there can be no influence of the shape of the boundary on the fracture process zone. It follows that G_f and c_f in an infinitely large specimen are independent of the specimen shape. Therefore, unambiguous definitions of G_f and c_f as fundamental material properties, independent of specimen size and shape, can be given as follows [1, 34, 36]: G_f and c_f are the energy required for crack growth and the elastically equivalent length of the fracture process zone, respectively, in an infinitely large specimen.

This definition of fracture energy can be mathematically stated as $G_f = \lim G_c = \lim(K_{Ic}^2/E')$ for $d \to \infty$, where G_c and K_{Ic} are the values of G and K_I (5), respectively, calculated for the measured peak load P_u and initial crack or notch length a_0 using linear elastic fracture mechanics. (More generally, one could define $G_f = \lim J$ for $d \to \infty$ where J is Rice's J-integral.)

For certain specimen geometries (for example, the center-notched panel loaded on the crack, the double cantilever specimen, the identation fracture test, the double-punch compression test and various chevron-notched specimens), $g'(\alpha)$ can be negative for some values of α and $g(\alpha)$ attains a minimum at a certain value, α_{min} [7–8, 15–17, 41–43]. In these cases, the peak load for ideally brittle materials (having no fracture process zone) occurs when α equals the value of α_0 or α_{min}, whichever is larger. When $g'(\alpha)$ is negative, the crack propagation is to a certain extent stable even under load control. The tests are, therefore, easier to control than the tests in which $g'(\alpha)$ is positive. These observations have lead various investigators to expect the toughness value obtained from the peak load of a test where $g'(\alpha_0)$ is initially negative to be size-independent, and therefore a true material property. Later, however, several investigators [42–43] observed that the fracture toughness values computed from this type of tests are size-dependent. The reason is that the equivalent relative crack length at the peak load depends strongly on the size, and is greater than both α_{min} and α_0. (It may be noted that, on the micro- or meso-scale, negativeness of $g'(\alpha)$ explains the observed stability of microcracking [44].)

In conventional testing, the apparent fracture energy, G_c, is usually determined by methods of linear elastic fracture mechanics without regard to the variations of the size of the fracture process zone, as if $\alpha = \alpha_0$ ($a = a_0$, $c = 0$) at failure. For that case, one gets (from (5)) $G_c = P_u^2 g(\alpha_0)/E'b^2 d$, which is size dependent. Substituting $P_u^2 = (\sigma_N bd/c_n)^2 = (Bf_u bd/c_n)^2 d_0/(d + d_0)$ and expressing Bf_u by using G_f from (9), one gets

$$G_c = \tau_N^2 \bar{d} = G_f \frac{d}{d + d_0} = G_f \frac{\bar{d}}{\bar{d} + c_f} = G_f \frac{\beta}{\beta + 1}. \tag{11}$$

The ratio $\beta_s = G_c/G_f = \beta/(1 + \beta)$, based on fracture energy [39], (as well as any other smooth monotonic function of β) could alternatively also be considered as a brittleness number; β_s varies from zero to one as the size increases from zero to infinity. According to (8), the size sensitivity of nominal strength [6] can be characterized by $(\Delta\sigma_N/\sigma_N)/(\Delta d/d) = -\beta_s/2$ or $\beta_s = -2(\Delta\sigma_N/\sigma_N)/(\Delta d/d)$.

Since $K_{Ic} = \sqrt{EG_c}$, the apparent fracture toughness is found [2] to vary as

$$K_{Ic} = \tau_N \bar{d}^{1/2} = K_{If}\left(\frac{\bar{d}}{\bar{d} + c_f}\right)^{1/2} = K_{If}\left(\frac{\beta}{1 + \beta}\right)^{1/2}. \tag{12}$$

These expressions follow from (6) and (7), and fracture toughness K_{If} is the value of K_{Ic} for an infinitely large specimen, corresponding to G_f, i.e. $K_{If} = \sqrt{E'G_f}$, which is a constant.

The size effect law has the advantage that its parameters G_f (or K_{If}) and c_f can be determined from the measured peak loads P_u by linear regression [3]. Algebraic rearrangement of (7) and (12) yields the linear plots

$$Y = AX + C \quad \text{or} \quad Y' = A'X' + C', \tag{13}$$

in which

$$X = \bar{d}, \quad Y = 1/\tau_N^2, \quad A = 1/K_{If}^2, \quad C = c_f/K_{If}^2, \tag{14}$$

$$X' = 1/\bar{d}, \quad Y' = 1/K_{Ic}^2, \quad A' = c_f/K_{If}^2, \quad C' = 1/K_{If}^2. \tag{15}$$

Each plot yields the slope of the regression line and its vertical intercept, from which the values of c_f and K_{If} follow. To make sure the size range of the specimens has been sufficient, one must use the well-known formulas of regression statistics for the coefficient of variation of the regression slope, and check whether its value is small enough [2].

3. Determination of *R*-curve and calculation of the load-displacement curve

The fracture process zone ahead of the notch is created by various mechanisms such as microcracking and bridging [37, 44–46]. The size of the fracture process zone grows as the fracture propagates [15, 23, 47]. This is manifested as an increase of the resistance $R(c)$ to fracture growth, representing the energy dissipated per unit length of fracture extension and unit width. As shown in [48], and refined in [34], the R-curve can be calculated from the material parameters obtained by size effect

$$R(c) = G_f \frac{g'(\alpha)}{g'(\alpha_0)} \frac{c}{c_f}, \tag{16}$$

in which

$$\frac{c}{c_f} = \frac{g'(\alpha_0)}{g(\alpha_0)}\left(\frac{g(\alpha)}{g'(\alpha)} - \alpha + \alpha_0\right). \tag{17}$$

These equations define the R-curve parametrically. After determining G_f and c_f from the size effect law, a series of α-values may be chosen and for each of them the length c of an elastically equivalent traction-free crack calculated from (17), and then $R(c)$ determined from (16). If c is specified, α may be solved by Newton iterations and subsequently $R(c)$ computed. The R-curve obtained according to (16) depends on specimen geometry.

For readers' convenience, the derivation of (16) and (17) is briefly as follows [34]. The energy balance at failure requires that $F(c, d) = G(\alpha, d) - R(c) = 0$ where $\alpha = a/d = \alpha_0 + c/d$. If we change the size slightly from d to $d + \delta d$ but keep the geometric shape (i.e. $\alpha_0 = $ constant), failure now occurs at $c + \delta c$, and since $G = R$ must hold also for $c + \delta c$, we must have $\partial F/\partial d = 0$. Geometrically, the condition $\partial F/\partial d = 0$ together with $F(c, d) = 0$ means that the R-curve is the envelope of the family of fracture equilibrium curves $F(c, d) = 0$ for various sizes d [48]. Because the R-curve is size-independent, we have $\partial R/\partial d = 0$ and so $\partial G/\partial d = 0$. Now we may substitute $P_u^2 = (\sigma_N bd/c_n)^2 = (Bf_u bd/c_n)^2/(1 + d/d_0)$ where $(Bf_u)^2 = c_n^2 E'G_f/d_0 g(\alpha_0)$ (according to (9)) into $G = P_u^2 g(\alpha)/E'b^2 d$ (5). We thus obtain for the critical states

$$G(\alpha, d) = G_f \frac{g(\alpha)}{g(\alpha_0)} \frac{d}{d + d_0}. \tag{18}$$

Substituting this into $\partial G/\partial d = 0$, differentiating, and noting that $\partial \alpha/\partial d = \partial \alpha_0/\partial d + \partial(c/d)/\partial d = -c/d^2 = -(\alpha - \alpha_0)/d$ (because $\partial \alpha_0/\partial d = 0$ for geometrically similar structures), we get

$$\beta = \frac{d}{d_0} = \frac{g(\alpha)}{(\alpha - \alpha_0)g'(\alpha)} - 1. \tag{19}$$

Furthermore, substituting this, along with the relations $(\alpha - \alpha_0)d = c$ and $d_0 = d/\beta = c_f g'(\alpha_0)$ (from (10b)) into (18), and setting $G(\alpha, d) = R(c)$, (16) and (17) are proven.

The dependence of the fracture toughness on c (i.e. the R-curve of fracture toughness) can be determined from the relation $K_{IR} = \sqrt{E'R}$:

$$K_{IR} = K_{If}\left(\frac{k(\alpha)k'(\alpha)c}{k(\alpha_0)k'(\alpha_0)c_f}\right)^{1/2}. \tag{20}$$

Equation (16) or (20) applies only as long as the fracture process zone grows and remains attached to the notch tip, which is approximately up to the peak load. The behavior for the post-peak regime will be discussed later.

The fact that the R-curves depend on the size of the specimen as well as its geometry (including the notch length) has been observed experimentally [31, 49–50]. It may also be noted that, according to some tests [50], interaction of the fracture process zone with the specimen boundary can even lead to a falling R-curve after a plateau is passed; however, the declining part of the R-curve is significant only for specimens with very small uncracked ligaments and usually affects only the final part of the post-peak response.

Knowing the R-curve, one can easily calculate the load-point displacement u_c due to fracture [31, 45, 50]. Let $u_0 = $ displacement calculated from elasticity as if there were no crack, and $u = u_c + u_0 = $ the total displacement. One determines first the total complementary energy, W_P, that would be released if the fracture occurred at constant load, P; from $\partial W_P/\partial a = bG = P^2 g(\alpha)/E'bd$, one gets

$$W_P = b \int_0^a G(a')\,\mathrm{d}a' = \frac{P^2}{E'b} \int_0^\alpha g(\alpha')\,\mathrm{d}\alpha'. \tag{21}$$

According to Castigliano's theorem,

$$u_c = \frac{\partial W_P}{\partial P} = \frac{2P}{E'b} \int_0^\alpha g(\alpha')\,\mathrm{d}\alpha'. \tag{22}$$

At the same time, from (5) for $G = R$:

$$P = b \sqrt{\frac{E'd}{g(\alpha)} R(c)}. \tag{23}$$

Choosing various values of α, one can calculate u_c and P from (22) and (23), which defines the load-deflection curve parametrically. As for u_0, it need not be calculated for our purpose since the elastic part of deformation dissipates no energy.

4. Calculation of the work-of-fracture from the *R*-curve

As we have seen, from the size effect law one can determine the *R*-curve, and from the *R*-curve one can calculate the diagram of the load vs. the load-point displacement. Thus, the work-of-fracture can be calculated from (1), or directly from the *R*-curve [16] (see Fig. 1c):

$$W_f = \int_0^l R(c)\,\mathrm{d}c. \tag{24}$$

Let now the *R*-curve, obtained as the envelope of fracture equilibrium curves of all sizes (and approximated by (16)–(17)), be called the master *R*-curve. For a finite size specimen, the master *R*-curve is followed only up to the peak load, i.e. through the regime in which the fracture process zone is growing while remaining attached to the notch tip (we of course consider only specimens for which $g'(\alpha) > 0$). Up to the peak load, the elastic zone surrounding the fracture process zone is opening up, because the effective stress intensity factor grows. After the peak load (beginning approximately but not exactly at the peak load), this factor ceases to grow, and the elastic zone around the fracture process zone ceases to open up further. Thus, after the peak load, the fracture process zone detaches itself from the notch tip and, as an approximate picture, travels forward without growing any more.

In view of this picture, the energy required for crack growth after the peak load must be kept constant and equal to the value of the master *R*-curve at the peak load [31], as shown in Fig. 1c. This has been verified by fitting of experimental load-deflection curves for rock [31] as well as high-strength concrete [32]. An important point is that the location of the point c_p at which

R becomes constant, i.e. the peak-load point, depends on specimen size d. Thus, the R-curve actually followed is different for each size and coincides with the master R-curve only during the load ascent up to the peak load.

Remark: The constancy of R after the peak load of course must cease to hold (and the R-value must start to decrease) as the crack front approaches the end of the ligament, because the boundary interference will tend to diminish the fracture process zone size. We exclude the terminal behavior from consideration and assume that it plays only a minor role.

The work-of-fracture may now be divided into two parts, W_1 and W_2, corresponding to the rising R-curve for the pre-peak regime, and to the horizontal portion for the post-peak regime (Fig. 1c):

$$W_f = W_1 + W_2, \tag{25}$$

in which

$$W_1 = b \int_0^{c_p} R(c)\,\mathrm{d}c, \quad W_2 = R(c_p)b(l - c_p). \tag{26}$$

For size d and at the peak load, one gets from (19) the effective relative crack length, α_p. From this, the effective length of the fracture process zone is $c_p = (\alpha_p - \alpha_0)d$. Then one can write

$$W_1 = b \int_{\alpha_m}^{\alpha_p} R(c)\frac{\mathrm{d}c}{\mathrm{d}\alpha'}\,\mathrm{d}\alpha', \tag{27}$$

in which from (17)

$$\frac{\mathrm{d}c}{\mathrm{d}\alpha} = -\frac{g'(\alpha_0)}{g(\alpha_0)}\frac{g(\alpha)\,g''(\alpha)}{[g'(\alpha)]^2} \tag{28}$$

and the value of α at $c = 0$ is given [34] by

$$\alpha_m = \alpha_0 + \frac{g(\alpha_m)}{g'(\alpha_m)}. \tag{29}$$

Finally

$$W_1 = bG_f c_f \frac{g'(\alpha_0)}{g(\alpha_0)}\bar{W}_1, \quad \bar{W}_1 = \frac{1}{g(\alpha_0)}\int_{\alpha_p}^{\alpha_m}\left[\frac{g(\alpha)}{g'(\alpha)} - (\alpha - \alpha_0)\right]\frac{g(\alpha)g''(\alpha)}{g'(\alpha)}\,\mathrm{d}\alpha, \tag{30}$$

$$W_2 = bG_f c_f \frac{g'(\alpha_0)}{g(\alpha_0)}\bar{W}_2, \quad \bar{W}_2 = \beta^2(1 - \alpha)(\alpha - \alpha_0)\frac{g'(\alpha)}{g(\alpha_0)}. \tag{31}$$

Thus, the fracture energy according to the work-of-fracture method used by RILEM is

$$G_f^R = \frac{W_1 + W_2}{bl} = G_f^{R_1} + G_f^{R_2},$$ (32)

in which

$$G_f^{R1} = \frac{\overline{W}_1}{\beta(1 - \alpha_0)} G_f, \quad G_f^{R2} = \frac{\overline{W}_2}{\beta(1 - \alpha_0)} G_f.$$ (33)

In the foregoing derivation we assumed that all the nonlinearity arises from the fracture process zone. This implies that G_f^P (which represents $\lim G_f^R$ for $d \to \infty$) is equal to G_f according to Bažant's definition [1]. Furthermore, in the foregoing calculation we tacitly assumed that the master R-curve, determined only from peak-load states, is also applicable, as an approximation, to fracture propagation during ascending load. This assumption is corroborated by the success in correctly predicting the measured load-deflection curves.

For large sizes, the entire master R-curve (16) can be used, which simplifies the calculations. In this case W_1 and $\overline{W}_1$ are constants, obtained by substituting $c_p = c_f$ and $\alpha_p = \alpha_0$, in (26), (27), and (30); also W_2 is a linear function of $l = d(1 - \alpha_0)$, i.e.

$$W_2 = b(l - c_f)G_f.$$ (34)

Finally, considering (2) and (25) one gets the size effect relation suggested by Planas, Elices and co-workers (3). The parameter l_p in (3) is related to c_f by

$$l_p = \left(1 - \frac{g'(\alpha_0)}{g(\alpha_0)} \overline{W}_m\right)c_f,$$ (35)

in which $\overline{W}_m(= \overline{W}$ for $\alpha_p = \alpha_0)$ is a geometry function calculated from (30), So we may conclude that (3) of Planas and Elices can be used, as an approximation, for $\beta > \beta_m(l > c_f)$, where

$$\beta_m = \frac{g(\alpha_0)}{g'(\alpha_0)(1 - \alpha_0)}.$$ (36)

The value of G_f^R which corresponds to β_m is

$$G_f^{Rm} = \frac{g'(\alpha_0)}{g(\alpha_0)} \overline{W}_m G_f.$$ (37)

But if (3) is used indiscriminately for $\beta < \beta_m$, discrepancy arises.

5. Size and shape dependence of fracture energy defined by work-of-fracture

Numerical results calculated from the foregoing equations for a three-point-bend fracture specimen with $\alpha_0 = \frac{1}{3}$ and the span-to-depth ratio $L/d = 2.5$ are plotted in Figs. 2–4 (the size is represented in terms of the nondimensional brittleness number, which is proportional to the beam depth). For this geometry, funtion $k(\alpha)$ was derived by finite element analysis and curve fitting [32] as

$$k(\alpha) = \frac{3L}{2d}\sqrt{\pi\alpha}(1-\alpha)^{-3/2}(1.0 - 2.50\alpha + 4.49\alpha^2 - 3.98\alpha^3 + 1.33\alpha^4). \tag{38}$$

Using this function, one can calculate $g(\alpha_0) = k^2(\alpha_0) = 14.20$, $g'(\alpha_0) = 2k(\alpha_0)k'(\alpha_0) = 72.71$, $\alpha_m = 0.4888$ (from (29)), $\bar{W}_m = 0.128$ (from (30) for $\alpha_p = \alpha_0$), $(l_p/c_f) = 0.343$ (from (35)), $\beta_m = 0.293$ (from (36)), and $G_f^{Rm}/G_f = 0.657$ (from (37).

Figure 2 shows the calculated ratios of the energy required for crack growth, i.e. the R-curve value from (16), and the apparent fracture energy G_c (11), both normalized by the (true) fracture energy, G_f, which represents the asymptotic value of the R-curve and is defined by extrapolation to infinite specimen size. The figure also shows the curve of the fracture energy according to RILEM work-of-fracture method, G_f^R, normalized to the true fracture energy, G_f (32).

An interesting observation may now be made: The RILEM fracture energy depends on the size of the specimen more than does the R-curve, while the opposite is desired from an ideal testing method.

Figures 3a and 3b show the curve of G_f^R/G_f for different relative notch depths, α_0, as function of $\bar{d}$ and d respectively. It is interesting to observe that, within a relatively broad range (Fig. 3a), namely $\frac{1}{3} \leqslant \alpha_0 \leqslant \frac{5}{6}$, the dependence of G_f^R on the intrinsic size $\bar{d}$ is nearly the same, while for a very short notch, such as $\alpha_0 = \frac{1}{6}$, this dependence becomes milder. On the other hand Fig. 3b shows strong dependence of G_f^R on the notch size for the same specimen size d.

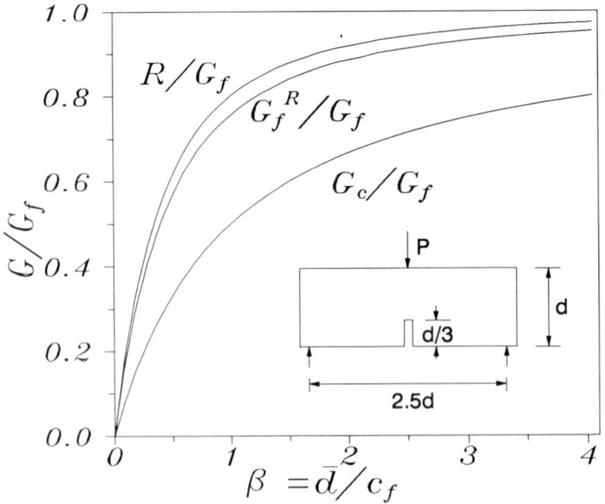

Fig. 2. Dependence of RILEM fracture energy G_f^R, relative to the true fracture energy G_f, on the specimen size characterized by brittleness number β, also the size dependence of the energy required for fracture growth, R, and apparent fracture energy, G_c, shown for comparison.

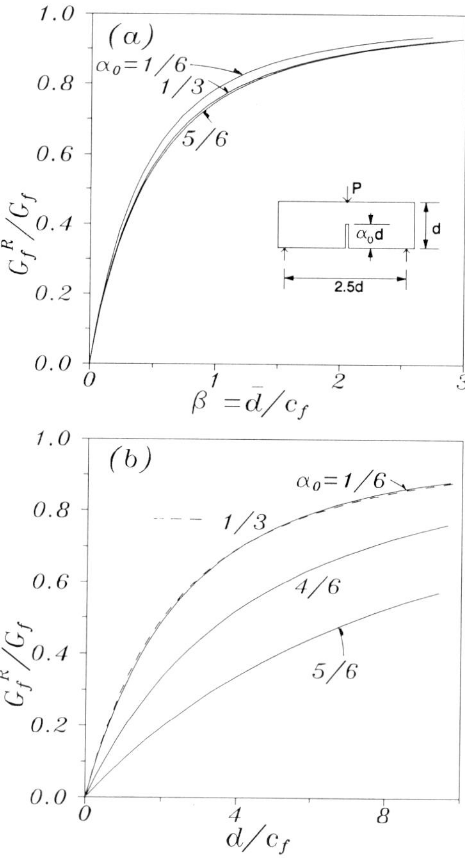

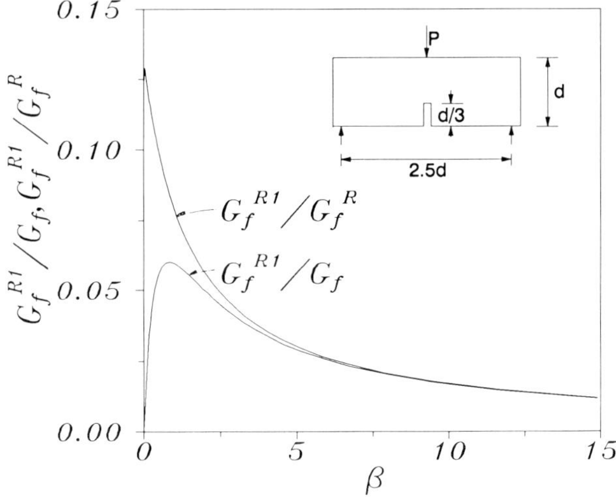

Fig. 3. Effect of relative notch depth $\alpha_0 = a_0/d$, on the value of RILEM fracture energy.

Fig. 4. Ratio of the pre-peak contribution to fracture energy, G_f^{R1}, to the RILEM fracture energy G_f^R and G_f.

Table 1. Analysis of test results of Maturana et al. [28]

T (°C)	E (GPa)	G_f^P (N/m)	G_f (N/m	κ_f
20	25.4	100	39	2.56
−10	27.2	227	87	2.61
−70	31.8	305	167	1.82
−170	47.9	332	316	1.05

It is of interest to determine the relative contributions of W_1 and W_2, (30) and (31), corresponding to the pre-peak and post-peak regimes. The pre-peak contribution of the work-of-fracture, G_f^{R1}, relative to the fracture energy, G_f, is plotted in Fig. 4 as a function of the specimen size, as characterized by the brittleness number β (33). We see that this ratio is generally rather small, always under 6 percent.

Figure 4 also shows the pre-peak contribution of the work-of-fracture relative to the fracture energy G_f^R according to the RILEM work-of-fracture method. This contribution decreases with increasing size from an initial value of about 13 percent.

The equivalence of G_f^P to G_f is true only theoretically, under the assumptions of the present analysis. The difference between G_f^P and G_f may be examined using the test results of Maturana, Elices and Planas [28] on saturated concrete at various low temperatures, presented in Table 1. The specimens have the same geometry as shown in Fig. 1, with the size range of $d = 50$–300 mm. The values of G_f^P are taken from [28–29] (calculated on the basis of (3)), and the values of G_f are estimated by nonlinear regression analysis based on (11) from the reported measured peak load values [28]. The moduli of elasticity used in the calculations of G_f are taken from uniaxial tension tests. As seen from the Table 1, G_f^P is much larger than G_f except for one case ($T = -170$°C). Figure 5 shows the dependence of G_f and G_f^P on temperature T, where a comparison is also made with the results from [51] on wet concrete at various elevated

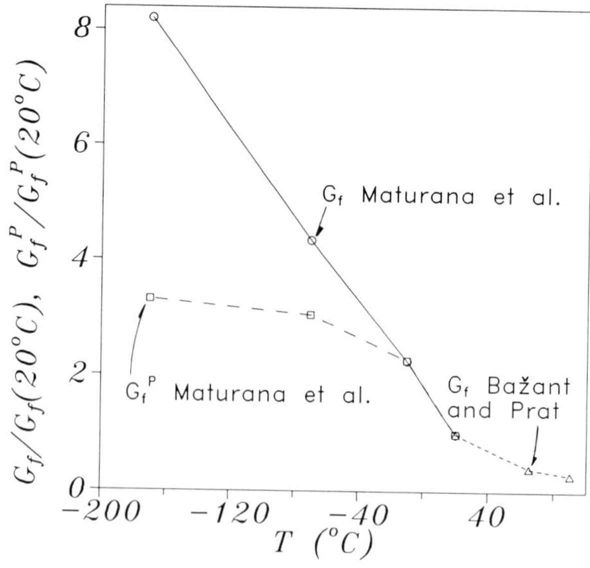

Fig. 5. Effect of temperature on fracture energy.

temperatures. The fracture energy values are normalized by the values corresponding to $T = 20°C$. It may also be of interest to note that the trend of G_f at low temperatures [28] is similar to that at high temperatures [51].

The ratio $\kappa = G_f^R/G_c$ indicates the degree of nonlinearity and ductility of the system [39], which again may depend on size and geometry. In the limit when $d \to \infty$, for which the nonlinearity due to fracture process zone is in theory eliminated, this ratio may be denoted as

$$\kappa_f = \frac{G_f^P}{G_f}. \tag{39}$$

If the extrapolation to infinity were exact, and if there were no dissipation nonlinearity sources outside the fracture process zone, one would have $\kappa_f = 1$. Practically, in most cases, $\kappa_f > 1$. The present theory predicts the RILEM fracture energy, G_f^R, to be always less than G_f in Bažant's definition. But this is contradicted by the results in Table 1 as well as many other places in the literature. This must be caused by some systematic errors, apart from random ones. Aside from a possible error in extrapolation to infinite size [36], one explanation might be that G_f^R is underestimated because dissipation sources outside the fracture process zone make a significant contribution to the total energy dissipation of the specimen. [6, 10, 21–22, 30, 52–53]. (For some geometries, e.g. simple unnotched tension specimens, this source of dissipation could yield a kind of size effect such that the apparent fracture energy based on the work-of-fracture increases with the specimen size [53].)

An important, spurious contribution to dissipation may come from the friction at the supports. As Planas and Elices [30] calculated, in order to make this contribution less than 1 percent of G_f^R, the friction coefficient of the rolling support in a three-point bend test would have to be below approximately 0.005L/d, which is not easy to achieve. For the wedge-splitting specimen with wedge angle 15°, the clamping force that the specimen exerts on the wedge requires the friction coefficient on the wedge to be below 0.0013 [30] if the friction contribution should be kept under 1 percent; this objective is next to impossible to attain in practice.

Figures 6 and 7 show variation of G_f^R with d (normalized by the G_f and d_0 values calculated from the size effect law) for a certain typical high strength concrete [32] and Indiana limestone [31], respectively. The data points are obtained from the area under the measured load-displacement diagram, (1) and (2). The curves in Figs. 6 and 7 show the variation of G_f^R predicted according to (32) from the observed dependence of peak nominal stress on the specimen size, (6) and (8). The comparisons are based on the modulus of elasticity obtained from the initial compliances of the fracture specimens rather than the standard tests on specimens. This yields mutually closer experimental (2) and predicted (32) values for G_f^R.

While the size effect method implies, for a given geometry, a unique pre-peak R-curve, because only one specimen size with the corresponding peak load corresponds to each $R(c)$-value, the R-curve to be used in the work-of-fracture method cannot be unique. The reason is that this method necessitates the R-values not only for all crack lengths but also for all specimen sizes; but the fracture process zone evolves probably differently in specimens of different notch lengths, not only in the post-peak regime but also in the pre-peak regime.

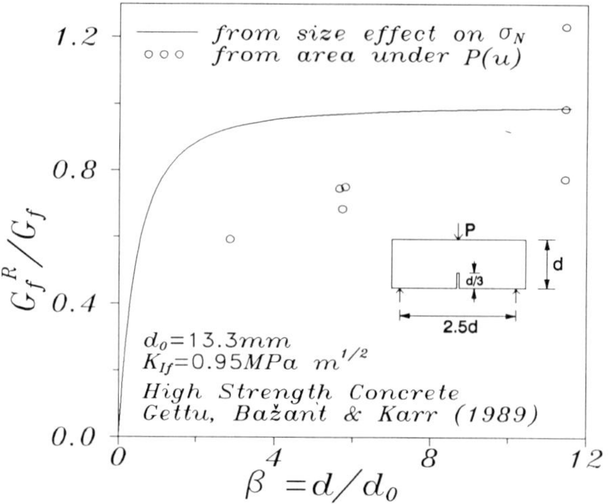

Fig. 6. Size dependence of G_f^R for high strength concrete [32].

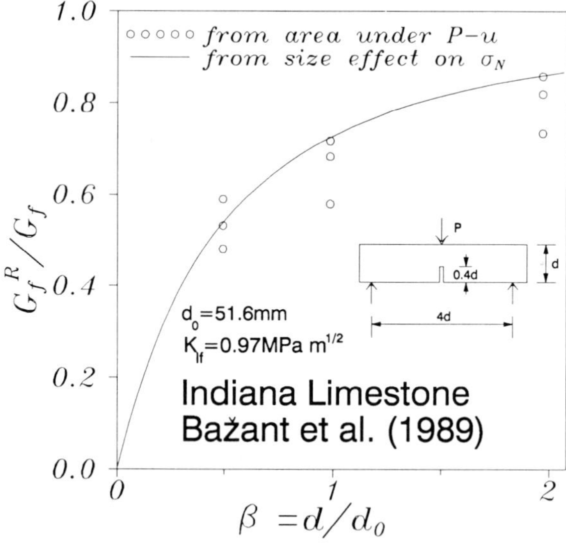

Fig. 7. Size dependence of G_f^R for Indiana limestone [31].

6. Conclusions

1. The basic premise of the present analysis is that although the size effect law of Bažant is approximate, it nevertheless has a broad range of validity, with an accuracy that is satisfactory compared to the inevitable scatter of test results. As shown in previous works, this law applies to the size range up to about 1:20, which suffices for most practical purposes. For a broader size range one would, of course, need a generalized size effect law, with additional parameters which would have to be geometry dependent; however, for the practical size range 1:20, such additional parameters are not needed, and the size effect law has the same form for various specimen shapes.

2. As a consequence of the aforementioned premise, the *R*-curve is geometry dependent, and may be calculated from size effect data on the maximum loads of fracture specimens of the same geometry. The *R*-curve can in turn be used to calculate the load-deflection curve and the work of fracture.

3. The calculations show that, in agreement with previous observations, the fracture energy defined by RILEM work-of-fracture method is not a constant but significantly increases with increasing specimen size. It is shown that the size dependence of the RILEM fracture energy is quite strong, in fact stronger than the size variation of the *R*-curve.

4. The calculations further show that (except for experimental scatter, dissipation outside the fracture process zone and other uncontrollable influences) extrapolation of the RILEM values of fracture energy to infinite specimen size must give exactly the same value of fracture energy for an infinitely large specimen as does the size effect method.

5. The contribution of the pre-peak work of fracture to the total fracture energy is rather small, and decreases with increasing specimen size.

6. As a by-product of the analysis, the fracture energy acording to the RILEM work-of-fracture method is seen to be sensitive to the relative notch depth in a three-point-bend test.

Acknowledgments

Partial financial support for the size effect theory of an *R*-curve has been obtained from the Center for Advanced Cement-Based Materials at Northwestern University (NSF grant DMR-8808432), and for the study of the RILEM method of determining the fracture energy from NSF Grant BSC-8818239 to Northwestern University. The previous research of the size effect in fracture was funded under AFOSR grant 91-0140 and contract F49620-87-C-0030DEF with Northwestern University. The second author wishes to express his thanks for a scholarship he received from Sharif University of Technology, Tehran, Iran.

References

1. Z.P. Bažant, in *Fracture of Concrete and Rock, Proceedings of SEM-RILEM International Conference*, Houston, June, 1987, S.P. Shah and S.E. Swartz (eds.), Springer-Verlag, NY (1989) 229–241; also Preprints, published by Society for Experimental Mechanics (1987) 390–402.
2. Z.P. Bažant and P.A. Pfeiffer, *ACI Material Journal* 84 (6) (1987) 468–480.
3. Z.P. Bažant, *Journal of Engineering Mechanics, ASCE* 110 (4) (1984) 518–35.
4. Z.P. Bažant, in *Proceedings of U.S.–Japan Seminar*, Tokyo, 1985, *Finite Element Analysis of Reinforced Concrete Structures*, C. Meyer and H. Okamura (eds.), ASCE, New York (1986) 121–150.
5. RILEM, *Materials and Structures* 18 (106) (1985) 285–290.
6. A. Hillerborg, *Materials and Structures* 18 (106) (1985) 291–96.
7. J. Nakayama, *Journal of American Ceramics Society* 48 (11) (1965) 583–87.
8. H.G. Tattersall and G. Tappin, *Journal of Material Science* 1 (3) (1966) 296–301.
9. A. Hillerborg, *Materials and Structures* 18 (107) (1986) 407–13.
10. P.E. Petersson, *Cement and Concrete Research* 10 (1) (1980) 78–101.
11. F. Moavenzadeh and R. Kuguel, *Journal of Materials* 4 (3) (1969) 497–519.
12. F. Radjy, *Cement and Concrete Research* 3 (4) (1973) 343–361.
13. A. Hillerborg, M. Modéer and P.-E. Petersson, *Cement and Concrete Research* 6 (6) (1976) 773–782.
14. M.P. Wnuk, *Journal of Applied Mechanics* 41 (1) series E (1974) 234–42.
15. A. Bascoul, F. Kharchi and J.C. Maso, in *Fracture of Concrete and Rock, Proceedings of SEM-RILEM International Conference*, Houston, June 1987, S.P Shah and S.E. Swartz (eds.) Springer-Verlag, NY (1989) 396–408.
16. M. Sakai, K. Urashima and M. Inagaki, *Journal of American Ceramics Society* 66 (12) (1983) 868–74.

17. T.B. Troczynski and P.S. Nicholson, *Journal of American Ceramics Society* 70 (2) (1987) 78–85.
18. F.H. Wittmann, K. Rokugo, E. Brühwiler, H. Mihashi and P. Simonin, *Materials and Structures* 21 (121) (1988) 21–32.
19. E. Brühwiler and F.H. Wittmann, *Engineering Fracture Mechanics* 35 (1/2/3) (1990) 117–25.
20. F.H. Wittmann, H. Mihashi and N. Nomura, *Engineering Fracture Mechanics* 35 (1/2/3) (1990) 107–15.
21. W. Brameshuber and H.K. Hilsdorf, *Engineering Fracture Mechanics* 35 (1/2/3) (1990) 95–106.
22. P. Nallathambi and B.L. Karihaloo, *Cement and Concrete Research* 16 (3) (1986) 373–82.
23. S.E. Swartz and T.M.E. Refai, in *Fracture of Concrete and Rock, Proceedings of SEM-RILEM International Conference*, Houston, June 1987, S.P. Shah and S.E. Swartz (ed.), Springer-Verlag, NY (1989) 242–254.
24. Y.S. Jenq and S.P. Shah, *Engineering Fracture Mechanics* 21 (5) (1985) 1055–1069.
25. S. XU and G. Zhao, in *Fracture Toughness and Fracture Energy, Test Methods for Concrete and Rock, Preprints of the Proceedings of an International Workshop*, Sendai, Japan, October, 1988, M. Izumi (ed.), Tohoku University, Sendai, Japan (1988) 48–62.
26. R.W. Davidge and G. Tappin, *Journal of Material Science* 3 (2) (1968) 165–73.
27. A. Hillerborg, in *Fracture Toughness and Fracture Energy, Test Methods for Concrete and Rock, Preprints of the Proceedings of an International Workshop*, Sendai, Japan, October 1988, M. Izumi (ed.), Tohoku University, Sendai, Japan (1988) 121–127.
28. P. Maturana, J. Planas and M. Elices, *Engineering Fracture Mechanics* 35 (4/5) (1990) 827–834.
29. J. Planas, P. Maturana, G. Guinea and M. Elices, in *Advances in Fracture Research, Proceedings of an International Conference (ICF7, Vol. 2)*, Houston, March 1989, Salama et al (eds.), Pergamon Press, Oxford (1989) 1890–1817.
30. J. Planas and M. Elices, in *Fracture Toughness and Fracture Energy, Test Methods for Concrete and Rock, Preprints of the Proceedings of an International Workshop*, Sendai, Japan, October, 1988, M. Izumi (ed.), Tohoku University, Sendai, Japan (1988) 1–18.
31. Z.P. Bažant, R. Gettu and M.T. Kazemi, *International Journal of Rock Mechanics and Mining Sciences* 28 (1991) 43–51.
32. R. Gettu, Z.P. Bažant and M.E. Karr, Fracture Properties and Brittleness of High Strength Concrete, Report No. 89-10/B627f, Center for Advanced Cement-Based Materials, Northwestern University (1989); also *ACI Materials Journal*, in press.
33. H. Tada, P.C. Paris and G.R. Irwin, *The Stress Analysis of Cracks Handbook*, 2nd. ed., Paris Production, St. Louis (1985).
34. Z.P. Bažant and M.T. Kazemi, *International Journal of Fracture* 44 (1990) 111–131.
35. Z.P. Bažant, and M.T. Kazemi, *Journal of American Ceramics Society* 73 (1990) 1841–1853.
36. J. Planas and M. Elices, in *Cracking and Damage, Strain Localization and Size Effect*, Proceedings of France-US Workshop, Cachan, France, 1988. J. Mazars and Z.P. Bažant (eds.), Elsevier, London (1989) 462–476.
37. H. Horii, Z. Shi and S.-X. Gong, in *Cracking and Damage, Strain Localization and Size Effect*, Proceedings of France-US Workshop, Cachan, France, 1988, J. Mazars and Z.P. Bažant (eds.), Elsevier, London (1989) 104–115.
38. Z.P. Bažant, and Y. Xi, Statistical Size Effect in Quasibrittle Structures: II. Nonlocal Theory, Report No. 90-5/616s(II), Center for Advanced Cement-Based Materials, Northwestern University, Evanston, Ill., 1990; also *Journal of Engineering Mechanics, ASCE*, in press.
39. J. Homeny, T. Darroudi and R.C. Bradt, *Journal of American Ceramics Society* 63 (5–6) (1980) 326–31.
40. A. Carpinteri, *International Journal of Solids and Structures* 25 (4) (1989) 407–429.
41. P. Marti, *ACI Materials Journal* 86 (6) (1989) 597–601.
42. J.L. Shannon, Jr. and D.G. Munz, in *Chevron-Notched Specimens: Testing and Stress Analysis*, ASTM STP 855, J.H. Underwood et al. (eds.), ASTM, Philadelphia (1984) 270–280.
43. K. Matsuki, in *Fracture Toughness and Fracture Energy, Test Methods for Concrete and Rock, Preprints of the Proceedings of an International Workshop*, Sendai, Japan, October, 1988, M. Izumi (ed.), Tohoku University, Sendai, Japan (1988) 234–248.
44. J. Huang, and C. Li, *Composites* 20 (4) (1989) 361–378.
45. Z.P. Bažant, *Cement and Concrete Research* 17 (6) (1987) 951–967.
46. M. Ortiz, *International Journal of Solids and Structures* 24 (3) (1988) 231–250.
47. F. de Larrad, C. Boulay and P. Rossi, in *Utilization of High Strength Concrete*, Proceedings of a Symposium, Stavanger, Norway, June, 1987, I. Holand et al. (eds.), Tapir Publishers (1987) 215–223.
48. Z.P. Bažant, J.-K. Kim and P.A. Pfeiffer, *Journal of Structural Engineering* 112 (2) (1986) 289–307.
49. Z.P. Bažant, S.-G. Lee and P.A. Pfeiffer, *Engineering Fracture Mechanics* 26 (1) (1987) 45–57.
50. M. Sakai and M. Inagaki, *Journal of American Ceramic Society* 72 (3) (1989) 388–394.
51. Z.P. Bažant and P.C. Prat, *ACI Materials Journal* 85 (4) (1988) 262–271.
52. L.J. Malvar and G.E. Warren, *Materials and Structures* 20 (120) (1987) 440–447.
53. M. Elices and J. Planas, in *Fracture Mechanics of Concrete Structures*, RILEM TC90-FMA, L. Elfgren (ed.), Chapman and Hall, London (1989) 16–66.

International Journal of Fracture **51**: 139–157, 1991.
Z.P. Bažant (ed.), *Current Trends in Concrete Fracture Research.*
© 1991 *Kluwer Academic Publishers. Printed in the Netherlands.*

Nonlinear fracture of cohesive materials

JAIME PLANAS and MANUEL ELICES
*Departamento de Ciencia de Materiales, Escuela de Ingenieros de Caminos, Universidad Politécnica de Madrid,
Ciudad Universitaria, 28040 Madrid, Spain*

Received 1 June 1990; accepted 1 November 1990

Abstract. The cohesive crack is a useful model for describing a wide range of physical situations from polymers and ceramics to fiber and particle composite materials. When the cohesive zone length is of the order of the specimen size, the influence method – based on finite elements – may be used to solve the fracture problem. Here a brief outline of an enhanced algorithm for this method is given. For very large specimen sizes, an asymptotic analysis developed by the authors allows an accurate treatment of the cohesive zone and provides a powerful framework for theoretical developments. Some recent results for the zeroth order and first order asymptotic approaches are discussed, particularly the effective crack concept and the maximum load size effect. These methods are used to analyze the effect of the size and of the shape of the softening curve on the value at the peak load of several variables for three point bent notched beams. The results show, among other things, that for intermediate and very large sizes the size effect curves depend strongly on the shape of the softening curve, and that only the simultaneous use of asymptotic and influence methods may give an adequate estimate of the size effect in the intermediate range.

1. Introduction

Since the introduction in the early sixties of particular cases of cohesive cracks by Barenblatt and Dugdale [1, 2], much work has been done in which the fracture of many different materials, from polymers to concrete and from ceramics to composites, is modelled as a cohesive crack.

The cohesive crack ability to model very different physical situations induces a certain dispersion in the methods of approach, depending partly upon the previous tradition of research in a particular field, usually identified by the material under study, partly by the requirements imposed by the material, in particular by the relative size of the fracture zone appearing in practice, which may span several orders of magnitude. In this paper we present a unified description of the nonlinear treatment of cohesive models for the simplest, yet important, case of mode I crack propagation, general enough to cover any softening behaviour and to span all the possible size ranges, from small sizes (large fracture zone) where limit analysis holds, to very large sizes (very small fracture zone) where Linear Elastic Fracture Mechancis (LEFM) is applicable [3, 4].

We start, in Section 2, with a brief look into the physical situations where implicitly or explicitly this kind of formulation has been used, and with the explicit macroscopic definition and general properties of cohesive cracks as they are understood in the remainder of the paper.

In the third section we present the details of an enhanced influence method for the numerical analysis of mode I cracking of symmetric specimens in the small size range. It is a very fast method, useful in research where the analysis of a large number of cases may be required. Its limitation is the element size of the underlying finite element mesh, which has to be a fraction of the fracture process zone size, thus limiting its practical application to small sizes.

The fourth section deals with the other extreme, and presents a formulation valid when the fracture zone is very small compared to all specimen dimensions. The essentials of the asymptotic analysis developed by the authors are explained for the zeroth and first order approximations. A new, simpler derivation of the central integral equation is given for the zeroth order approximation. which is the equivalent of the *small scale yielding* approach in classic fracture mechanics. The basic results for the first order asymptotic approach are presented without full demonstrations, the emphasis being put on the interpretation of the results, particularly on the effective crack concept, and on the equation for the size effect.

The fifth section applies the above methods to the study of the effect of specimen size on several variables at the peak load: the maximum load itself, the initial crack tip opening displacement, the size of the cohesive zone and the J integral. This study is performed for three point bent single edge notched specimens, and the influence of the shape of the softening curve is investigated by considering rectangular (Dugdale), linear, and quasi-exponential softening. A summary of the essential conclusions – among which the importance of complementing asymptotic and small size analyses with each other – closes the paper.

2. Cracking of cohesive materials

2.1. *The different underlying physics*

The cohesive crack model is an idealized approximation of a physical localized fracture zone, adopted with the primary objective of mathematical simplicity while preserving the most relevant aspects of physical reality. The essence of this model is the description of non-linearity by means of a relationship between cohesive stresses and crack openings, and its great virtue is that, with the adequate relationship, it can be used to describe a wide range of physical situations.

During the sixties, the cohesive crack was used to model two different behaviors: brittle fracture was analyzed by Barenblatt [1], and ductile fracture by Dugdale [2]. It is well known that for brittle materials, the Griffith and Barenblatt approaches are equivalent (see, for example, [5]). In the Barenblatt formulation the singularity is abolished by means of a highly localized distribution of cohesive forces at the crack tip; these forces have a molecular origin and derive from a potential. Barenblatt limits the analysis to cases where the size of the cohesive zone is very small with respect to the dimensions of the specimen and does not need to postulate the shape of the force-displacement relationship, but only its integral (the fracture surface energy, as shown below).

Dugdale put forward his celebrated model to represent plastic yielding in slender zones at the end of cracks, and postulated a uniform cohesive stress on a suitable cohesive zone defined in such a way that the stress singularity was also removed.

Cohesive cracks have been used to model cracking in polymers and ceramic materials. Crazes are modelled with cohesive stresses representing the forces transmitted by the fibrils at the craze-bulk polymer interface [6]. In non-transforming ceramics, for example, a grain-bridging model of crack resistance appears to be common to a wide range of materials, specially non-cubic ceramics that fail by intergranular failure, and, again, cracking is modelled by a cohesive crack [7].

Cracking of brittle matrix composites, whether particle or fiber reinforced, is another example. The influence of the bridging particles or fibers is to restrain the opening of the crack and consequently cause a reduction of stresses in the matrix ahead of the crack tip. The increased toughness of reinforced brittle matrix composites comes from this fiber bridging or aggregate pull-out. This effect is modelled by a cohesive crack. Note that in this kind of material, two types of cohesive forces may be considered: a highly localized distribution at the crack tip, to model the brittle matrix, and a wider distribution to account for the reinforcement effect.

Concrete may be considered a particular case of particulate brittle matrix composite, but the development of fracture models for it has followed a different history. The application of cohesive models to concrete was pioneered by Hillerborg and coworkers around 1976 [8, 9]. Their approach was macroscopic and based on the results of a stable tensile test. Since then, much work has been devoted to finding numerical and experimental methods to study and predict concrete fracture. These methods may be useful for the analysis of other materials, provided the adequate material functions are used. A microscopic foundation of the cohesive stress-crack opening curve has been put forward by Bazant and Ortiz, based on microcracking [10, 11], and by Horii, based on microcracking at the early stages and frictional aggregate bridging at the later stages [12, 13]. In the remainder of this section we focus on the macroscopic formulation of general cohesive crack models.

2.2. General macroscopic properties of cohesive crack models

An extensive review of cohesive crack models can be found in a recent RILEM report [14], so only the most relevant aspects are summarized here, and only cohesive crack models with no bulk dissipation are considered. The characterization of a cohesive crack model includes the definition of the bulk behaviour, the specification of the condition for crack formation and the specification of the equations for crack evolution. In practice, such conditions are restricted to

1. *Bulk behaviour*: The solid is homogeneous and the behaviour at any point is isotropic linear elastic – with Young's modulus E and Poisson's ratio v – as long as the major principal stress does not reach a critical value, the tensile strength f_t.
2. *Crack initiation*: When the maximum principal stress reaches f_t, fracture is initiated and strain localization takes place in what is called the fracture process zone (FPZ). The FPZ is modelled as a cohesive crack where the strain localization is idealized as a displacement jump or crack opening, while cohesive stresses simulate the softening behaviour.
3. *Crack propagation*: Once the cohesive crack has formed, the stress transferred through the crack faces is assumed to depend upon the relative displacement of the crack faces as:

$$\sigma = f(w) \quad \text{with } f(0) = f_t \quad \text{and } f(w) \geqslant 0, \tag{2.1}$$

where $f(w)$ describes the softening behaviour of the material and is called the *softening curve* and is considered a material function. Coplanar, monotonic crack growth is assumed. In practice, the values of $f(w)$ are assumed to be zero for crack openings exceeding a *critical crack opening* w_c.

The work needed to open monotonically a crack of unit surface, the *specific work supply*, is also a material function given by

$$W_F(w) = \int_0^w f(w')\,dw'.$$

(2.2)

The specific work supply needed to fully open a unit surface of crack is a material property called the *specific fracture energy* G_F (or simply, fracture energy), i.e.

$$G_F = \int_0^\infty f(w')\,dw' = \int_0^{wc} f(w')\,dw'.$$

(2.3)

From the material parameters, E, v, f_t, and G_F, two independent parameters having the dimension of length were introduced; the first was called *characteristic length* by Hillerborg and defined as

$$l_{ch} = \frac{G_F E}{f_t^2},$$

(2.4)

where, from now on, E is understood as the generalized Young's modulus. The second parameter called *characteristic crack opening*, is defined as

$$w_{ch} = \frac{G_F}{f_t}.$$

(2.5)

It is useful to write the softening equation $f(w)$ in dimensionless form by defining dimensionless stresses σ^* and crack openings w^* as

$$\sigma^* = \sigma/f_t, \quad w^* = w/w_{ch}.$$

(2.6)

Equation (2.1) may then be written as

$$\sigma^* = F(w^*), \quad \text{where } F(w^*) = f(w_{ch}w^*)/f_t.$$

(2.7)

Once the softening function is given, the mode I crack propagation analysis may be performed in a conceptually straightforward manner. Figure 1 sketches a precracked specimen loaded in monotonic mode I, in which a cohesive zone of length R has developed in front of the initial crack. The problem may be treated as a linear elastic one with variable and nonlinear boundary conditions. The easier formulation is the crack-length controlled one, in which the size of the process zone R is taken as the independent variable. In the usual case of proportional external loading the load factor is determined simultaneously with the stresses and crack openings in the cohesive zone by imposing standard conditions on the initial specimen boundary, and forcing (2.1) to be satisfied on the cohesive zone, together with the essential requirement that the stress be less than or equal to f_t everywhere around the cohesive zone, which enforces the stress

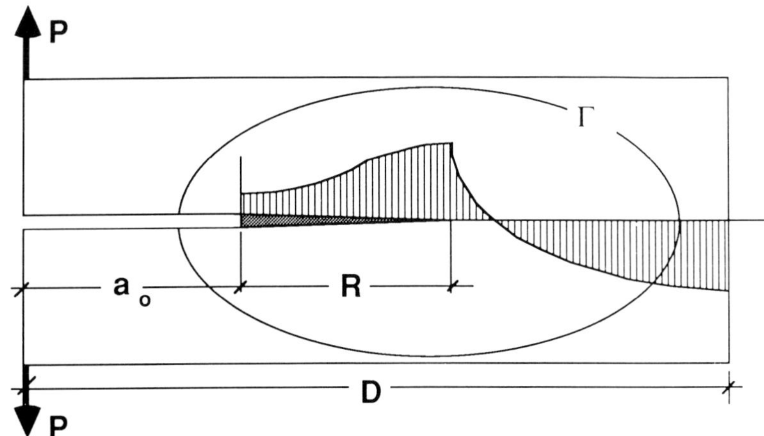

Fig. 1. Mode I cohesive crack growth in a symmetric specimen.

solution near the cohesive crack tip to be non-singular, or, equivalently, the 'true' stress intensity factor to be zero.

There are different methods of treating this problem, some of which will be dealt with in the next sections. However, there is one essential feature common to all the cohesive crack problems as defined here. The J integral along paths surrounding the cohesive zone, such as path Γ in Fig. 1, accepts a closed form expression depending on the $CTOD$, or crack opening at the initial crack tip location. Such expression is obtained by computing J along a limiting path defined by the crack (real plus cohesive) faces themselves, which is possible because the actual solution is non-singular. A straightforward calculation yields the result

$$J = \int_0^{CTOD} f(w')\,dw' = W_F(CTOD),\tag{2.8}$$

where the second equality follows from (2.2).

Apart from the above universal equation, the choice of the method of analysis depends essentially on the size of the specimen. More precisely, it depends on the size of the specimen relative to the FPZ size, which in turn depends on the material. It will be seen that the size must be compared to the material characteristic size l_{ch} defined by (2.3), which may range from some tens of microns for brittle ceramics to some tens of centimeters for concrete and fiber reinforced materials – a span of seven orders of magnitude. In the next sections we analyze some of the existing methods separating those suitable for small specimens from those for very large specimens.

3. The analysis of a cohesive crack in mode I for small size specimens

In problems where the cohesive zone R is of the order of the size of the specimen, the finite element method or some special adaptation of it may be used to solve the problem. In a recent RILEM report these methods are reviewed in detail by Hillerborg and Rots [15]. Here we limit ourselves to a special purpose method very well suited for research, where in many instances the crack path is known in advance and a large number of runs on the same geometry may be needed for analysis of different models. We call this method the influence method, which was first used for concrete fracture analysis by Hillerborg and coworkers [8, 9]. Our starting point is the same as theirs, and only the enhanced method of solution and the ability to handle non-polygonal equations make a difference.

The method consists of using the superposition scheme shown in Fig. 2. The real situation is written as a sum of linear elastic solutions that are obtained by standard finite element computations for an auxiliary specimen with a crack longer than the maximum expected cohesive crack in the actual situation. The central equation derived from Fig. 2 relates the crack opening at each node to the corresponding closing nodal forces and external load by means of the influence matrices K and C as

$$w_i = C_i P - K_{ij} p_j, \quad i, j = 1,...N, \tag{3.1}$$

where N is the number of node-pairs along the crack in the auxiliary specimen.

Using standard dimensional analysis, it is obvious that (3.1) may be written in the form

$$w_i = C_i^* D \frac{\sigma_N}{E} - K_{ij}^* D \frac{\sigma_{cj}}{E}, \quad i, j = 1,...N, \tag{3.2}$$

where D is a characteristic dimension of the specimen, σ_N is a *nominal* stress defined as P/BD (P is the load and BD a characteristic cross-sectional area), and σ_{cj} the nodal closing stresses, and the starred influence matrices are now dimensionless and size independent, and may be obtained directly once and for all by running a standard finite element program for a specimen having unit thickness, unit size and unit Young's modulus, for the $N + 1$ cases implicit in Fig. 2.

Using the dimensionless variables defined in (2.6), and taking into account the boundary conditions along the initial crack, the cohesive zone and the remaining intact ligament, we find

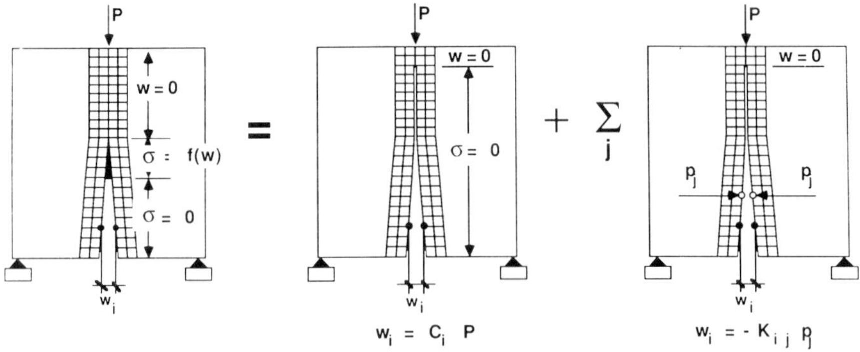

Fig. 2. Superposition scheme at the basis of the influence method.

the following set of $2N + 1$ equations:

$$w_i^* = C_i^* D^* \sigma_N^* - K_{ij}^* D^* \sigma_{cj}^*, \quad i, j = 1,..., N, \tag{3.3}$$

$$\sigma_{ci}^* = 0, \qquad\qquad i = 1,..., N_{\text{notch}}, \tag{3.4}$$

$$\sigma_{ci}^* = F(w_i^*), \qquad\qquad i = 1 + N_{\text{notch}},..., 1 + N_{\text{notch}} + N_{\text{cohes}}, \tag{3.5}$$

$$w_i^* = 0, \qquad\qquad i = 1 + N_{\text{notch}} + N_{\text{cohes}},..., N, \tag{3.6}$$

where N_{notch} is the number of node-pairs on the initial notch, and N_{cohes} the number of cracked nodes, which varies from 0 to $N - N_{\text{notch}} - 1$ and is the independent variable. Note that there is an index overlap in (3.5) and (3.6) which forces the node-pair at the cohesive crack tip to be in an incipient cracking situation: $w^* = 0$ and $\sigma^* = 1$. The dimensionless size D^* is given by

$$D^* = D/l_{ch}. \tag{3.7}$$

In previous approaches [9], these equations were directly solved numerically for σ_N^*, w_i^*, and σ_{cj}^*. In the approach used by the authors, the number of previously used equations is reduced from $2N + 1$ to N_{cohes} according to the following algorithm: we define primed vectors and matrices on the initial ligament as:

$$V_i' = V_{i + N_{\text{notch}}}^*, \quad Q_{ij}' = Q_{i + N_{\text{notch}}\, j + N_{\text{notch}}}^*. \tag{3.8}$$

With this convention we substitute (3.4) into (3.3) and find, for the node-pairs over the initial ligament:

$$w_i' = C_i' D^* \sigma_N^* - K_{ij}' D^* \sigma_{cj}', \quad i, j = 1,..., N_{\text{lig}}, \tag{3.9}$$

where $N_{\text{lig}} = N - N_{\text{notch}}$. We numerically invert $\boldsymbol{K'}$, only once for one given notch-do-depth ratio, to find

$$\sigma_{ci}' = L_i' \sigma_N^* - M_{ij}' w_j'/D^*, \quad i, j = 1,..., N_{\text{lig}}, \tag{3.10}$$

where $\boldsymbol{M'} = \boldsymbol{K'}^{-1}$ and $\boldsymbol{L'} = \boldsymbol{M'C'}$ are determined once and for all for a given initial notch depth. We next substitute (3.6) and (3.5) to get the final set of $1 + N_{\text{cohe}}$ nonlinear equations:

$$F(w_i') = L_i' \sigma_N^* - M_{ij}' w_j'/D^*, \quad i, j = 1,..., 1 + N_{\text{cohe}}, \tag{3.11}$$

$$w_i' = 0, \qquad\qquad i = 1 + N_{\text{cohe}}. \tag{3.12}$$

These equations are easily solved using a Newton-Raphson method with the softening function analytically defined, and the convergence is much faster than when the initial set of equations (3.3)–(3.6) is used without modification. The time involved in inverting $\boldsymbol{K'}$ becomes negligible when a great number of cases (different sizes and softening functions) are analyzed.

The authors use this method on a desktop computer HP9826 with the total influence matrices for 36 node-pairs obtained by Petersson in Lund University and the average time required for the complete analysis of one case is about 1 minute. The analysis of one case with 90 nodes using a MicroVax II with influence matrices obtained using ANSYS takes approximately half this time [16]. In this paper all the results are obtained with the 36 node-pair program.

This method of analysis fails for large sizes as a simple examination of the problem shows. Indeed in this procedure the size of the elements grows in proportion to the specimen size and, in the limit, the element itself will be much larger than the cohesive zone, hence unable to describe the stress and crack opening variations inside this zone. Indeed, (3.11) shows that for large dimensionless size $D*$ the second term in the second member vanishes and the maximum load is obtained at the first step, for vanishing cohesive zone, at a constant nominal stress given by

$$\sigma^*_{\text{Nmax apparent}} = 1/L'_1. \tag{3.13}$$

Unfortunately this value is totally element-size-dependent and vanishes as the square root of h/D where h is the size of the element, as will be shown elsewhere. In conclusion, this method may be used for sizes where the cohesive zone is larger than the element size. Until further analyses quantify the question, one may estimate that at least 4 or 5 elements must fit into the cohesive zone in the range of interest to get a good description of the overall behavior.

For very large sizes a special asymptotic analysis was developed by the authors which allows a very accurate treatment of the cohesive zone, and, even more important, provides a powerful framework for theoretical developments, i.e. for proving theorems. This analysis is summarized in the next section.

4. The analysis of a cohesive crack in mode I for very large sizes

The solution of the cohesive crack problem for infinite size $(D \gg R)$ may be based on known elastic solution for a family of loadings using a superposition scheme, or Green's formulation. The difference in the approaches comes from the Green's function chosen.

In their asymptotic analysis the authors adopted as the basic solution that for an elastic crack of variable length in a body of any size D, and then expressed the full solution as a power series of R/D, from which different order approximations were obtained [17, 18, 19]. This original deduction is quite lengthy and here we adopt a shorter and easier derivation based on an integral transformation of a classic formulation. In doing this we lose a great deal of deductive power since higher order approaches cannot be derived. We gain, however, in clarity and brevity.

4.1. Infinite size: the zeroth order asymptotic approach

We consider a body with an initial crack in front of which a relatively small cohesive zone has formed. The zeroth order approximation is obtained when all terms of order R/D or higher are neglected. It is formally equivalent to substituting the real fields by those corresponding to a semi-infinite crack in an infinite body. Following the essence of Horii's approach [12], we look for a solution as a superposition of the three cases shown in Fig. 3. The first case corresponds

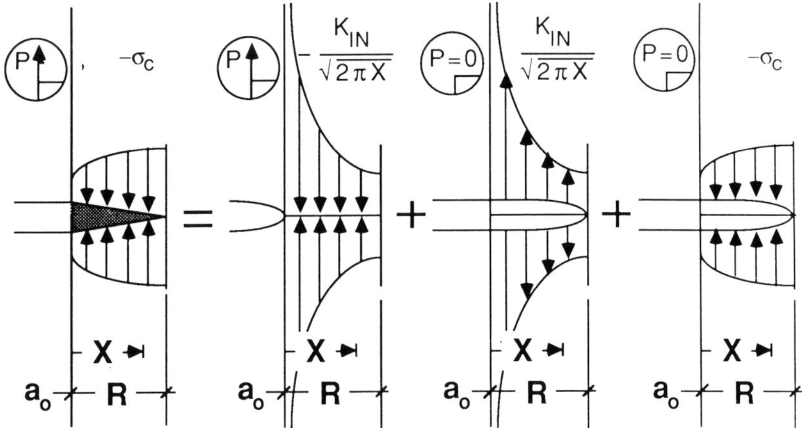

Fig. 3. Superposition scheme for the zeroth order asymptotic approach.

to the elastic loading of the initial crack, of length a_0, which leads to a stress distribution ahead of the crack tip that may be written, as long as $X \leqslant R \ll D$, as

$$\sigma = -\frac{K_{IN}}{\sqrt{2\pi X}} + \text{negligible terms,} \tag{4.1}$$

where the minus sign indicates that these stresses are of the closing type. This first case does not contribute to the stress intensity factor at the cohesive crack tip or to the crack openings on the fracture zone, but it does contribute to the crack openings over the initial crack. The second case corresponds to a crack of length $a_0 + R$ loaded on its faces by opening stresses opposite to those in (4.1). The third case correspond to the closing cohesive stresses σ_c. The second and third case may be unified, and the corresponding stress intensity factor at the cohesive tip, and the crack openings written in terms of the solution for a point load on the faces of a semi-infinite crack in an infinite body, as depicted in Fig. 4. The resulting integral equations may be written, using the reduced coordinate $x = X/R$, as follows:

$$K_I = K_{IN} - \sqrt{\frac{2R}{\pi}} \int_0^1 \sigma_c(x)(1-x)^{-1/2}\, dx, \tag{4.2}$$

$$w(x \leqslant 1) = \frac{8RK_{IN}}{\sqrt{2\pi E}}(1-x)^{1/2} + \frac{4R}{\pi E}\int_0^1 \sigma_c(x')\ln\frac{\sqrt{1-x}+\sqrt{1-x'}}{|\sqrt{1-x}-\sqrt{1-x'}|}\, dx'. \tag{4.3}$$

These equations are general elastic solutions that have to be used in conjunction with the material-related equations

$$K_I = 0, \tag{4.4}$$

$$\sigma_c(x) = f[w(x)] \quad \text{for } 0 < x \leqslant 1, \tag{4.5}$$

where (4.4) forces the resultant stress field to be non-singular, and (4.5) is the replica of the softening curve (2.1) which has to hold on the cohesive zone. Of course the resultant set of

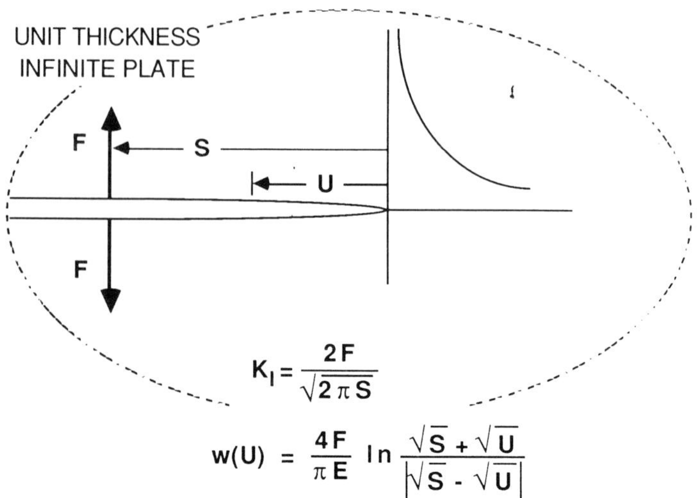

$$K_I = \frac{2F}{\sqrt{2\pi S}}$$

$$w(U) = \frac{4F}{\pi E} \ln \frac{\sqrt{S} + \sqrt{U}}{|\sqrt{S} - \sqrt{U}|}$$

Fig. 4. Elastic solutions for a semi-infinite crack in an infinite medium.

equations may be directly used to look for a solution, as was done for example by Horii [12, 13]. But the equation is rather formidable and includes a logarithmic kernel which is somewhat difficult to handle numerically and analytically.

We obtain the author's asymptotic formulation by introducing a new density function $k(x)$, zero everywhere except on the cohesive zone, defined through the following integral relationship of the Volterra type:

$$\sigma_c(x) = \int_0^x k(u)(x - u)^{-1/2}\,du, \tag{4.6}$$

which may be inverted to give:

$$k(x) = \frac{\sigma_c(0^+)}{\pi\sqrt{x}} + \frac{1}{\pi}\int_0^x (x - x')^{-1/2}\,d\sigma_c(x'). \tag{4.7}$$

Substitution of (4.6) and (4.4) into (4.2) and (4.3) lead to double integral expressions which, after inverting the order of integration and performing the first of the integrals result in the following simpler integral equations:

$$K_{IN} = \sqrt{2\pi R}\int_0^1 k(u)\,du, \tag{4.8}$$

$$w(x) = \frac{8R}{E}\int_0^1 k(u)(u - x)^{1/2}H(u - x)\,du \quad \text{for } x \leqslant 1, \tag{4.9}$$

where $H(x)$ stands for the Heaviside step function.

The next step is to write (4.5). To gain in generality it is best to write this equation in dimensionless form using (2.6) and (2.7), together with the dimensionless cohesive zone size R^* and the dimensionless weight function $k^*(u)$:

$$R^* = \frac{R}{l_{ch}}, \qquad k^*(u) = \frac{k(u)}{f_t}. \tag{4.10}$$

With this, the governing integral equation becomes:

$$\int_0^x k^*(u)(x - u)^{-1/2}\, du - F\left(8R^* \int_x^1 k^*(u)(u - x)^{1/2}\, du \right) = 0. \tag{4.11}$$

This integral equation is nonlinear, but contains very simple Volterra's kernels which lead to triangular tangential stiffness matrices overlapping only along the diagonal, if a finite element scheme is used with simple point collocation algorithm. The only special feature to notice is that, as shown by (4.7), $k^*(u)$ is singular at the origin whenever the stress is discontinuous at the initial crack tip, which is generally the case. The easiest expedient to solve this problem is to use shape functions which already contain the $u^{-1/2}$ singularity. The authors use as unknown a function $g(u) = k^*(u)u^{1/2}$. They use 58 finite elements over (0, 1) with linear interpolation within the elements. Equation (4.11) is forced to be satisfied at the 59 nodes. All the integrals over the elements are well known binomial integrals and are analytically defined in a subprogramme. The softening function and its first derivative are analytically defined also, and a Newton-Raphson algorithm used to find the solution for a set of values of R^*. An automatic peak detection is used, based on the property that for an infinite body the peak is reached when $CTOD = w_C$.

To close this section, we emphasize that the equations above are specimen or geometry-independent. In the formulation above we based these equations on the known solutions for a semi-infinite crack in an infinite medium, but they were shown to hold for any geometry by taking the limit $R/D \to 0$ [17, 18, 19].

4.2. Quasi-infinite size: first order approach and the effective crack concept

The results of the preceding section gain in interest when a further order of approximation is introduced in the analysis. Indeed, it was shown that the structure of the far field for a cohesive crack model coincides with that of an *effective elastic crack* field to order $(R/D)^2$ [18, 19]. It is worth clarifying the strict sense of this statement as illustrated in Fig. 5. The left part of this figure represents the real structure, in which a cohesive zone of size R has developed monotonically in front of the initial crack while the load increased up to P. The right part of the figure represents an imaginary (virtual) specimen, purely elastic, subject to the same load P, but with a crack of length $a = a_0 + \Delta a_\infty$, where Δa_∞ is an effective crack extension, and subindex ∞ stands for the fact that we assume we are approaching the asymptotic limit of *infinite* size. The equivalence is stated in the following way: one may choose (calculate) Δa_∞ in such a way that the far fields (shaded zones in Fig. 5) coincide except for terms of the order $(R/D)^2$ or higher. The interesting point is that the value of Δa_∞ for each loading step may be computed from the solution of the zeroth order approximation, (4.11), as shown elsewhere

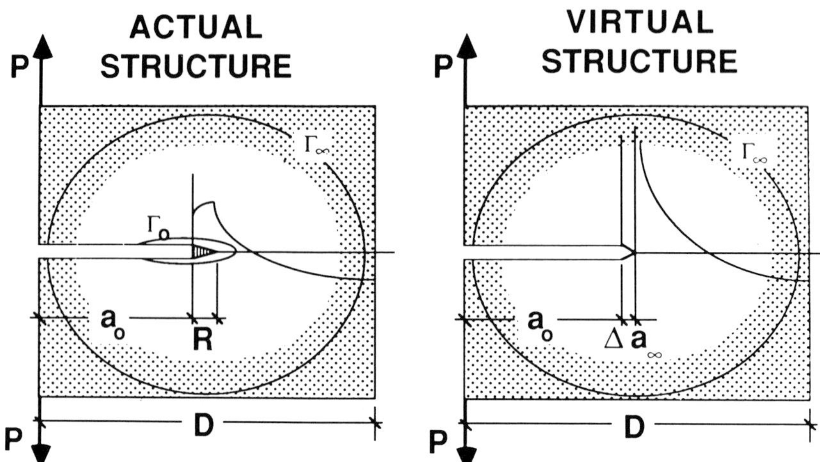

Fig. 5. Effective crack extension in the infinite size limit. The remote fields (shaded area) are the same to order R/D.

[18, 19]. The expression for the infinite size effective crack extension in terms of the solution for the density function $k^*(x)$ is given by

$$\Delta a_\infty = R \frac{\displaystyle\int_0^1 u k^*(u)\, du}{\displaystyle\int_0^1 k^*(u)\, du}, \qquad (4.12)$$

which is no more than the expression for the centroid of the density function k^* (or k).

The effective crack extension for quasi-infinite size is independent of geometry and size and its evolution depends only on the shape of the softening function. Its determination requires the solution of (4.11). When this solution is obtained, a maximum for the loading parameter K_{IN} is reached for a certain critical FPZ size R_C, for which a critical effective crack extension $\Delta a_{C\infty}$ may be computed. These two magnitudes take unique values for a given material, but depend strongly on the shape of the softening curve, as it will be shown later.

The existence of an effective crack equivalence, together with the closed form expression for the J integral, (2.8), may be exploited with profit. In the virtual specimen depicted in Fig. 5 there is a stress singularity at the crack tip with a stress intensity factor $K_{Ivirtual}(a)$ given by an equation of the well known form

$$K_{Ivirtual}(a) = \sigma_N \sqrt{D}\, S(a/D), \qquad (4.13)$$

where σ_N is a *nominal* stress defined as P/BD (P is the load and BD a characteristic cross-sectional area) and $S(a/D)$ is the geometrical shape factor. Note that σ_N is the same for the virtual and real specimens.

For the actual cohesive cracked body the stress intensity factor has no meaning as a measure of the stress singularity, because the stress is finite at the crack tip, but for comparative purposes we can define a *nominal* stress intensity factor as a loading parameter involving load level, size, and initial shape:

$$K_{\mathrm{IN}} := \sigma_N \sqrt{D}\, S(a_0/D). \tag{4.14}$$

The actual near-tip fields may be related to the virtual ones by using path-independent integrals. In particular we may use the J integral along a path completely surrounding the cohesive zone, as path Γ_0 in Fig. 5, which is equal to the J integral performed along a path Γ_∞ fully enclosed into the far field region (dashed region), which may in turn be approximated by performing the integral in the virtual specimen, then obtaining

$$J := J_{\mathrm{actual}} = J\{\Gamma_\infty\} \approx J_{\mathrm{virtual}}\{\Gamma_\infty\} = J_{\mathrm{virtual}} = \frac{1}{E}[K_{\mathrm{Ivirtual}}(a_0 + \Delta a_\infty)]^2. \tag{4.15}$$

To simplify this equation while retaining the same order of accuracy, one can use a two term expansion of (4.13) and definition (4.14) to obtain

$$J := J_{\mathrm{actual}} = \frac{K_{\mathrm{IN}}^2}{E}[1 + (2S_0'/S_0)\Delta a_\infty D^{-1}] + o(1/D), \tag{4.16}$$

where $S(a/D)$ is the shape factor introduced in (4.13), $S'(a/D)$ is its first derivative with respect to the argument and subindex 0 refers to initial crack length, and $o(1/D)$ stands for a function approaching zero faster than its argument.

When this expression is combined with (2.8) one gets the following relationship relating the load level, the effective crack extension and the *CTOD*:

$$\frac{K_{\mathrm{IN}}^2}{E}[1 + (2S_0'/S_0)\Delta a_\infty D^{-1}] = W_F(CTOD). \tag{4.17}$$

This equation is the basis for the determination of the size effect on the peak load for very large sizes, a subject dealt with in the next section.

4.3. Asymptotic size effect and related parameters

It is well known that the maximum nominal stress decreases as the size increases for a set of geometrically similar precracked bodies. Based on the dimensional analysis and on the limiting LEFM behaviour for large sizes, a fairly general expression for this size effect was found by the authors for cohesive crack models [20]. This expression takes the form of a power series expansion and may be written in terms of maximum nominal stress intensity factor as

$$EG_F(K_{\mathrm{INmax}})^{-2} = 1 + C_1 l_{ch}/D + C_2(l_{ch}/D)^2 + \cdots, \tag{4.18}$$

where l_{ch} is the characteristic length (2.4) and C_i are dimensionless coefficients that depend on geometrical ratios and on the slope of the softening curve. The structure of the first of these coefficients may be found from the previous asymptotic analysis for a wide range of geometries.

Indeed in (4.17) we may notice that the right hand member is a monotonic function reaching an absolute maximum (and a plateau) at some moment along the process. When one considers the first member, and the fact that the effective crack extension is also monotonically increasing, one realizes that the two following cases can occur:

1. If $S_0' < 0$ the nominal stress intensity factor is strictly monotonic to first order in $1/D$, and no maximum load is present to this order. This means that the position and intensity of the maximum is controlled by higher order terms. If $S_0' = 0$ the behaviour depends on higher order derivatives, and again, on higher order terms. This kind of behavior corresponds to what are called *non-positive geometries*, and the resulting size effect is termed *anomalous* and will not be treated further here.

2. If $S_0' > 0$ then a maximum of the load does exist to first order. This corresponds to what was termed *positive geometries* in [21], and to *normal* size effect behavior. This size effect may be obtained to first order accuracy by analysis of (4.17) around the maximum for the zeroth order approximation ($D^{-1} = 0$), which is obviously attained for $CTOD = w_C$ and, by definition, for $\Delta a_\infty = \Delta a_{C\infty}$. When first order terms are taken into account it appears that the maximum occurs near the previous one, the modifications of the point of occurrence giving rise to terms of order higher than the first. The resulting normal asymptotic size effect stems out then from (4.17) as

$$EG_F(K_{INmax})^{-2} = 1 + (2S_0'/S_0)\Delta a_{C\infty} D^{-1} + o(1/D). \tag{4.19}$$

From (4.19) it may be noticed that two cracked samples of the same material are asymptotically equivalent when the factor $(2S_0'/S_0)D^{-1}$ takes the same value for both structures. Consequently, we may define an intrinsic size D_i [21] as

$$D_i^{-1} = (2S_0'/S_0)D^{-1}. \tag{4.20}$$

At a first glance it may appear that D_i depends on the selection of D. But it may be easily shown that this is not so, because (4.20) is equivalent to the definition

$$D_i^{-1} = \left. \frac{\partial \ln K_1^2}{\partial a} \right|_{a = a_0}, \tag{4.21}$$

where in the partial derivative with respect to crack length a, the load and all remaining dimensions of the specimen remain constant. With this definition, (4.19) takes the simpler form

$$EG_F(K_{INmax})^{-2} = 1 + \Delta a_{C\infty}/D_i + o(1/D_i). \tag{4.22}$$

As pointed out in [21] this intrinsic size may become negative, which apparently is a serious drawback. However, it is essential to remember that this analysis was done only for positive geometries (i.e. when the stress intensity factor increases with crack length, or $S_0' > 0$), and that non-positive size would correspond to non-positive geometry, which corresponds to an anomalous size effect. Hence it seems that the negativeness of the intrinsic size corresponds to a physically and mathematically distinct situation and may not be incorrect *a priori*.

5. Application to size effect analyses

The methods previously analyzed may be used efficiently to study the effect of the size on several macroscopic variables. The size effect *par excellence* is that referring to maximum load, and was introduced as a useful research tool by Bazant in 1984 [3]. However, the effect of the size on any other variable may also be analyzed and this was done in the present research for variables related to the peak. These variables are the fracture process zone size R, the J integral, the crack tip opening displacement CTOD, and, of course, the maximum load.

The results for the three softening curves defined in Fig. 6 are presented, with the aim of showing the influence of the length of the softening 'tail', or w_C, on the size effects. The Dugdale softening is of academic and historic interest, the linear softening has been extensively used as the simplest representation of gradual softening, and the quasi-exponential softening is a more realistic approximation for the softening behavior of concrete.

Table 1 shows the results for the different variables at the peak for the infinite size limit obtained by solving (4.11). The results for the Dugdale model are all exact, and some of the

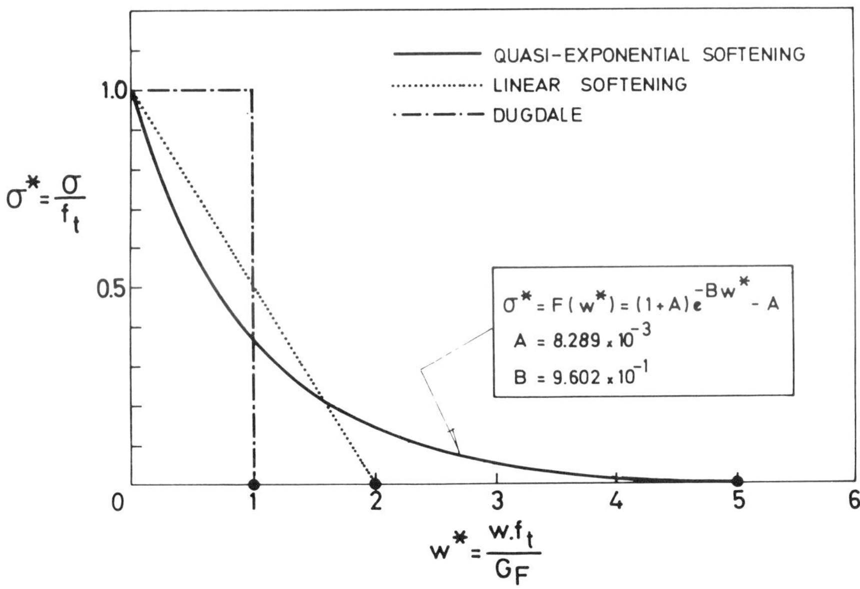

Fig. 6. Definition of the softening curves.

Table 1. Results for infinite size

Model	$(K_{INmax})^2$	$\Delta a_{C\infty}$	$R_{peak\,\infty}$	$CTOD_{peak\,\infty}$	$J_{peak\,\infty}$
Dugdale	EG_F	$\dfrac{\pi}{24}l_{ch}$	$\dfrac{\pi}{8}l_{ch}$	w_{ch}	G_F
Linear softening	EG_F	$0.419l_{ch}^{\dagger}$	$0.731l_{ch}^{\dagger}$	$2w_{ch}$	G_F
Quasi-exponential softening	EG_F	$2.48l_{ch}^{\dagger}$	$2.92l_{ch}^{\dagger}$	$5w_{ch}$	G_F

†Approximate; see text.

results for the other models are also exact. The approximate values are indicated by a dagger, and they are thought to have more than 3 digits accuracy. We remark that these results are geometry-independent.

The size dependence of the different variables was analyzed for notched beams tested in three point bending, having a span-to-depth ratio of 4 and an initial notch-to-depth ratio of 0.5. Figures 7 to 10 show the results represented versus the inverse of the intrinsic size previously defined, a plot that has the advantage of bringing the infinite size into focus.

Figure 7 is a dimensionless plot of the peak load expressed in terms of the inverse of the square of the stress intensity factor. This kind of plot has the advantage of resulting in very simple lines for some classical models and has been used repeatedly by the authors [21, 22]. In particular, Bazant's size effect law [3] takes the form of a straight line in this plot. The most important aspect shown by Fig. 7 is that the Dugdale model gives a result very close to Bazant's law, and that the deviation from a straight line is larger the longer the tail of the softening curve. The influence method may give accurate results up to intrinsic sizes about one material characteristic size (heavy lines), but if only these small size results were known, the extrapolation to larger sizes could not be confidently performed as is clear from the figures. Knowledge of the asymptotic solution transforms extrapolation into interpolation and then the behavior in the intermediate size range may be reasonably estimated as the light dashed curves shown in the figures.

For the remaining three variables, the differences found in the intermediate size range are even more striking. Figures 8 and 9 clearly show that for softening functions with a long tail, both the size of the cohesive zone at peak load, R_{peak}, and the crack tip opening at the peak load, $CTOD_{\text{peak}}$, vary very fast in the intermediate range, where accurate solutions are lacking, and interpolation is difficult. Needless to say, extrapolation from small size computations to the intermediate and large size would be highly inaccurate. For the J integral, J_{peak}, Fig. 10 shows

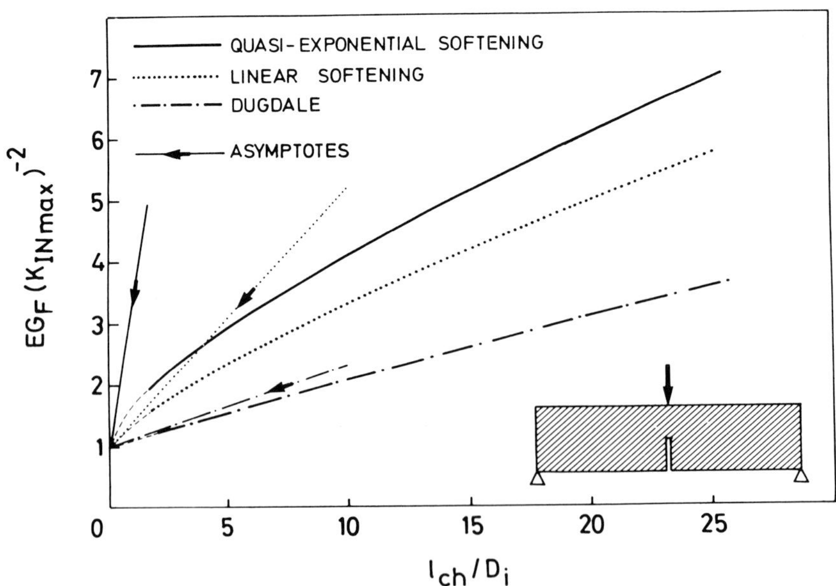

Fig. 7. Size effect plot for the maximum load – showing first order asymptotes – for the three softening curves defined in Fig. 6.

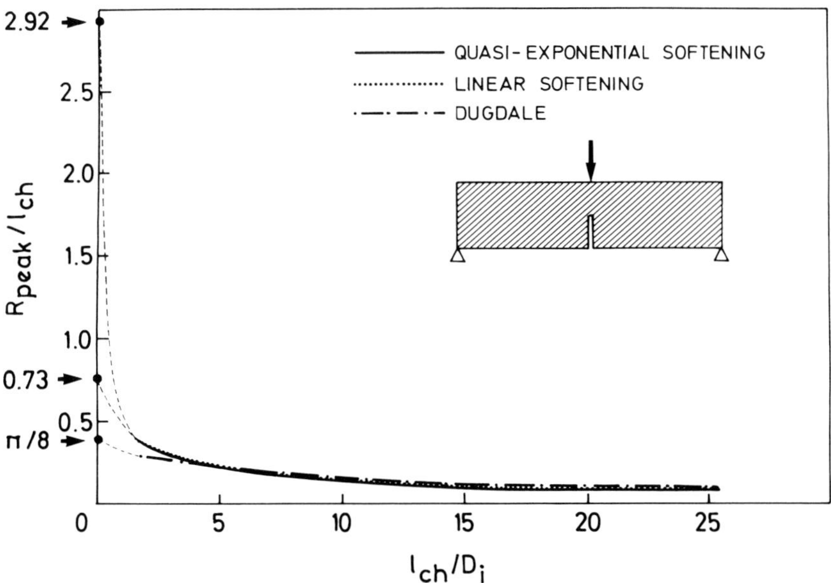

Fig. 8. Variation of the cohesive zone length at peak load with the size of the specimen for the three softening curves defined in Fig. 6.

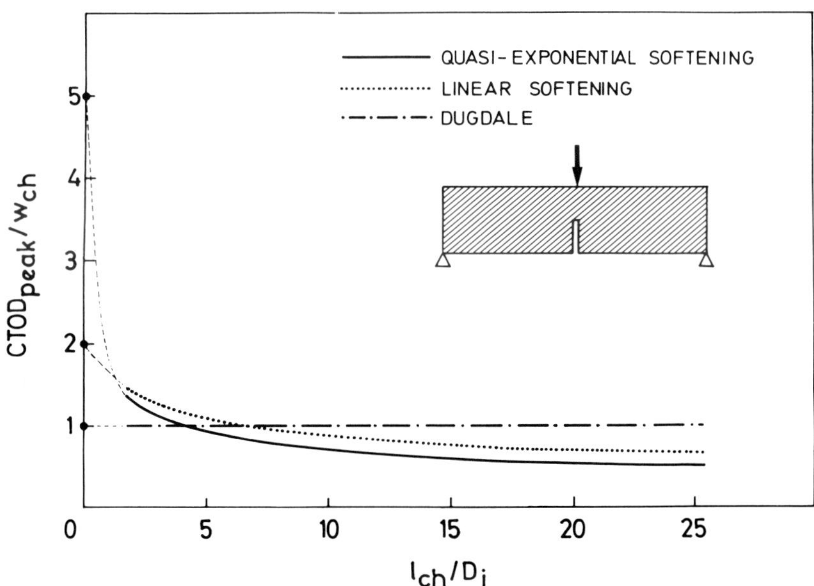

Fig. 9. Variation of the CTOD at peak load with the size of the specimen for the three softening curves defined in Fig. 6.

that the interpolation is of intermediate difficulty, because it may be shown that J must approach the limit of large sizes with horizontal tangent, and hopefully interpolation could be done within a ± 5 percent accuracy.

Asymptotic analysis is, then, a useful tool to allow interpolation in the intermediate size range, where the influence method cannot give accurate information at a reasonable cost, i.e. without increasing tremendously the number of elements of the underlying finite element mesh. To get

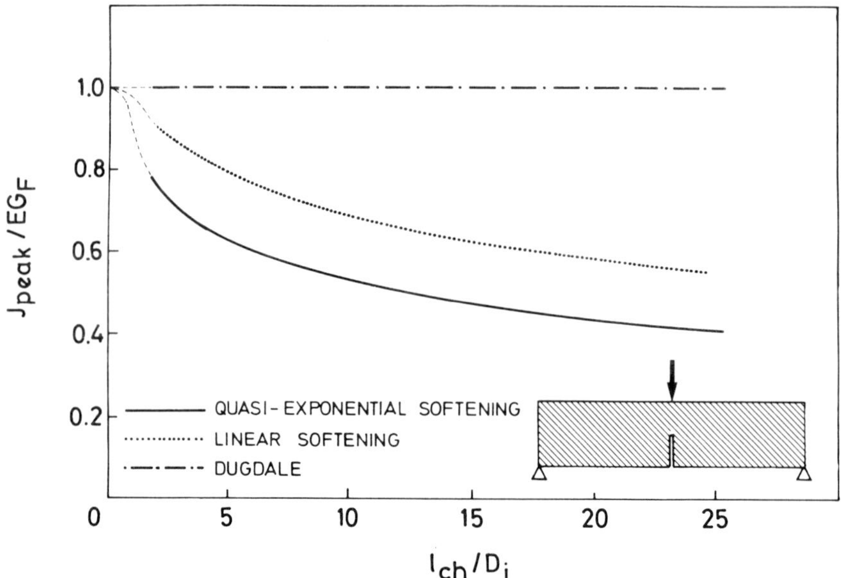

Fig. 10. Variation of the J integral at peak load with the size of the specimen for the three softening curves defined in Fig. 6.

accurate solutions in the intermediate range, a special purpose method of analysis should be developed. Higher order asymptotic approximations following the method developed by the authors may be a reasonable solution. Work is in progress to analyze this possibility.

6. Conclusions

From the previous analyses, the following conclusions may be drawn.

1. Cohesive cracks are simplified useful models to represent fracture of various materials.
2. For small specimens, the modified version of the influence method here presented is a powerful tool for research purposes.
3. The need of keeping the size of the finite element smaller than the fracture process zone precludes the use of the influence method for very large sizes.
4. Direct extrapolation of the small size results, without further information, may lead to large errors for large sizes.
5. The asymptotic analysis presented in Section 4 provides an adequate framework for computational as well as theoretical treatment of cohesive cracks for very large sizes.
6. Using the inverse of the intrinsic size as the abscissa generates optimum plots to represent the size effect for positive geometries when large sizes are considered.
7. The results of the application of the above methods to notched beams show that the length of the 'tail' of the softening curve – the dimensionless critical crack opening – largely influences the size effect trend in the intermediate and large size range (intrinsic size larger than about 0.5 or 1 times the characteristic size, or abscissa in the plots *less* than 1 or 2).

8. For some variables the interpolation in the intermediate size range seems to be achievable with low relative error (maximum load, J at peak). For other variables the variation is too fast and too large for long tailed softening curves, and accurate interpolation is difficult (CTOD and FPZ at peak).

9. Work is needed to provide an efficient method of analysis over this intermediate size range. Higher order asymptotic approaches following the method derived by the authors might prove useful in doing so.

Acknowledgements

The authors gratefully acknowledge Comisión Interministerial de Ciencia y Tecnología, Spain, for providing financial support for this research under grants PA85-0092 and PB86-0494.

References

1. G.I. Barenblatt, *Journal of Applied Mathematics and Mechanics* 23 (1959) 622–636.
2. D.S. Dugdale, *Journal of Mechanics and Physics of Solids* 8 (1960) 100–104.
3. Z.P. Bazant, *Journal of Engineering Mechanics, ASCE* 110 (1984) 518–538.
4. Z.P. Bazant, *Applied Mechanics Reviews* 39 No 5 (1986) 675–705.
5. J.N. Goodier, in *Fracture, an Advanced Treatise, Vol. 2*, H. Liebowitz (ed.), Academic Press, New York (1968) 1–66.
6. T. Ungsuwarungsri and W.G. Knauss, *ASME Journal of Applied Mechanics* 55 (1988) 44–58.
7. S.J. Bennison and B.R. Lawn, *Acta Metallurgica* 37 No 10 (1989) 2659–2671.
8. A. Hillerborg, M. Modeer and P.E. Petersson, *Cement and Concrete Research* 6 (1976) 773–782.
9. P.E. Petersson, Crack Growth and Development of Fracture Zones in Plain Concrete and Similar Materials, Report TVBM 1006, University of Lund, Sweden (1981).
10. Z.P. Bazant, *Cement and Concrete Research* 17 (1987) 951–967.
11. M. Ortiz, *International Journal of Solids and Structures* 24 (1988) 231–250.
12. H. Horii, A. Hasenaga and F. Nishino, *Preprints, SEM-RILEM International Conference on Fracture of Concrete and Rock*, Society for Experimental Mechanics, Bethel CT 6801, USA (1987) 299–307. Also in *Fracture of Concrete and Rock*. S.P. Shah and S.E. Swarts (eds.), Springer Verlag, New York (1989) 205–219.
13. H. Horii, S. Zihai and S-X Gong, in *Cracking and Damage, Strain Localization and Size Effect*, J. Mazars and Z.P. Bazant (eds.), Elsevier Applied Science, London (1989) 104–115.
14. M. Elices and J. Planas, in *Fracture Mechanics of Concrete Structures: From Theory to Applications*, L. Elfgren (ed.), Chapman and Hall, London (1989) 16–66.
15. A. Hillerborg and J. Rots, in *Fracture Mechanics of Concrete Structures: From Theory to Applications*, L. Elfgren, ed., Chapman and Hall, London (1989) 128–146.
16. J.L. Lorca, J. Planas and M. Elices, in *Fracture of Concrete and Rock: Recent Developments*. S.P. Shah, S.E. Swartz and B. Barr (eds.), Elsevier Applied Science, London (1989) 357–368.
17. J. Planas and M. Elices, *Anales de Mecánica de la Fractura* 3 (1986) 219–227.
18. J. Planas and M. Elices, A new method of asymptotic analysis of the development of a cohesive crack in mode I loading. Report 87–02 Depto. de Ciencia de Materiales, Universidad Politécnica de Madrid, Spain (1987).
19. J. Planas and M. Elices, *International Journal of Fracture* (1992).
20. J. Planas and M. Elices, in *Fracture Toughness and Fracture Energy of Concrete*, F.H. Wittmann (ed.), Elsevier Science, Amsterdam (1986) 381–390.
21. J. Planas and M. Elices, in *Cracking and Damage, Strain Localization and Size Effect*. J. Mazars and Z.P. Bazant (eds.), Elsevier Applied Science, London (1989) 462–476.
22. J. Planas and M. Elices, *Engineering Fracture Mechanics* 35 No 1/2/3 (1990) 87–94.

International Journal of Fracture **51**: 159–173, 1991.
Z.P. Bažant (ed.), Current Trends in Concrete Fracture Research.
© 1991 *Kluwer Academic Publishers. Printed in the Netherlands.*

Size effect and continuous damage in cementitious materials

J. MAZARS[1], G. PIJAUDIER-CABOT[1] and C. SAOURIDIS[2]
[1]*Laboratoire de Mécanique et Technologie, E.N.S. de Cachan/C.N.R.S./Université Paris 6, 94235 Cachan Cedex, France;*
[2]*Engineering System International, 20 rue Saarinen, Silic 270, 94578 Rungis Cedex, France*

Received 1 June 1990; accepted 1 November 1990

Abstract. The dependence between the mechanical properties of concrete structures and their sizes is a problem addressed in this paper. Two types of size effect are distinguished: the first one is probabilistic and is related to random distributions of defects in a volume of material; the second one is purely deterministic and is related to fracture propagation in brittle heterogeneous media. These two aspects are combined in a continuous damage model. The initiation of damage is probabilistic and the evolution of damage is nonlocal. Comparisons with experiments on notched and unnotched bending beams show that the model provides a realistic prediction of the various size effects.

1. Introduction

Cementitious materials and ceramics often exhibit size effect, that is a dependence of the mechanical properties of the material on the size of the structure. For example L'Hermite reported in 1973[1] that the nominal tensile strength of concrete deduced from three point bending specimens decreases when the volume of the specimen increases. Kadlecek et al. [2] also noted that the tensile strength was different depending on the type of loading applied to specimens of the same volume and geometry. Experiments show that the average tensile strength computed from four point bending beams is increased by 200 percent and by 220 percent in three point bending from what is measured in direct tension. Experiments on notched specimens of homothetic sizes made of concrete, rocks and ceramics by Bazant and coworkers [3, 4, 5] also demonstrate that the nominal tensile and shear strengths depend on the size of the specimens.

All these experiments show two key points:

1. The tensile strength of cementitious materials and ceramics as predicted by elasticity or limit analysis is a function of the volume of the specimen and of the stress field inside the structure. Part of this size effect can be attributed to the existence of initial defects in the material before any stress is applied to it. The amount of defects is a function of the volume of material tested and their topological distribution inside the structure must certainly be related to the stress field applied to it. For example, chances are that the most important defect is not located in the critical mid-span cross-section of a specimen loaded in three point bending. Thus its influence on the tensile strength is much less than in direct tension where every cross-section is critical. In this paper, we will show that this type of size effect may be handled by a probabilistic description of damage initiation.

2. The fracture toughness measured from notched specimens also varies with the size of the structure. Failure in these materials is often the result of progressive microcracking in a fracture process zone whose size is related to the size of inhomogeneities or aggregates. The fracture process zone lying ahead of a crack loaded in mode I has approximately a width of three times the maximum size of the aggregates and a length of five to six times the size of the aggregates [6]. The volume of this process zone is constant whereas the size of the structure may be changed. Thus, the stress and energy redistribution between the process

zone and the rest of the structure is important for small enough specimens. For large structures the size of the process zone is negligible and damage is concentrated in a relatively small zone around the crack tip. Bažant [3] has proposed an approximate formula for the prediction of this size effect which combines limit analysis for small specimens and linear elastic fracture mechanics for very large specimens:

$$\sigma_N = Bf_t[1 + (d/d_a\lambda)]^{-1/2}, \tag{1}$$

where σ_N is the nominal tensile stress at failure for the considered specimen or structure, f_t is the tensile strength; d is the width of the specimen (all specimens are homothetic), d_a is a reference length function of the aggregate size; B and λ are constants. A typical result is shown in Fig. 1 for three point bending beams. We may notice that for usual structures the prediction of the maximum strength lies between the two criteria on a smooth curve. This type of size effect has to be related to the evolution of damage prior to failure and not to the existence of initial defects since in these tests the notch may be regarded as the largest defect inside the material and its size remains constant with regard to the size of the specimen tested.

In Part 3 of this paper we will show that continuous damage models predict this size effect in a purely deterministic fashion. Finally, numerical applications and comparisons with experiments on unnotched and notched specimens will demonstrate the usefulness of the combination of probabilistic and deterministic size effects.

2. General framework – continuous damage mechanics

Failure in cementitious materials and ceramics is mainly due to progressive damage in the form of microcrack propagation and void growth. On a macroscale the effect of microcracks and

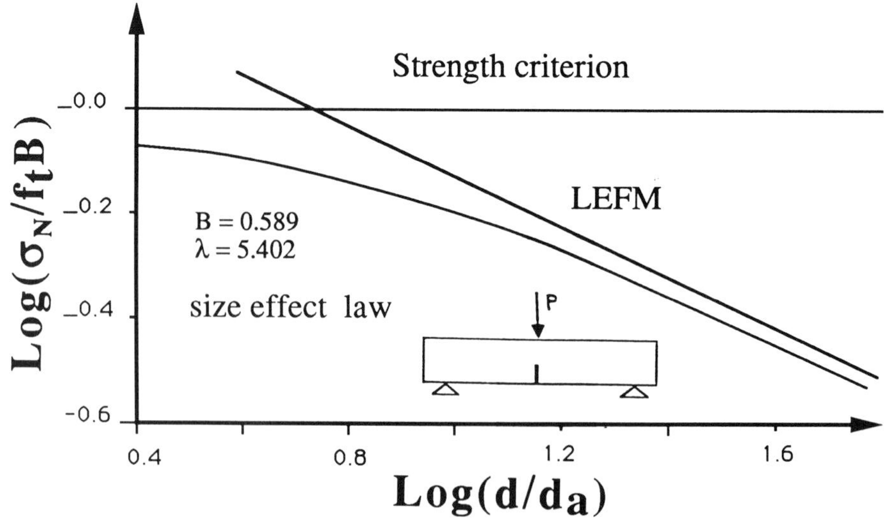

Fig. 1. Size effect on concrete in three point bending (after [3]).

voids may be combined into a single internal state variable called damage. The evolution of damage has various consequences on the response of the material.

1. The effective cross-section of the material diminishes due to voids and microcracks and the stiffness of the material decreases. Moreover, if the microcracks are oriented in a specific direction the material exhibits a damage-induced anisotropy.

2. Microcrack or void growth also produces irreversible strains. If the surface of the crack is not smooth it may not close completey under zero stress. Void growth produces also a change of mass density and irreversible variations of volume. Finally, the redistribution of internal stresses due to hydration of cement plays also a pivotal role in this phenomenon. There is a possibility for crack closure only if the material is subjected to cyclic loading. It was experimentally demonstrated that the material recovers also its initial stiffness when the load changes from tension to compression [7].

3. Damage creates also a potentiality for internal friction. The microcracks, once they are created, may be loaded in mode II and III. In this case, friction between the surfaces of microcracks produces "plastic " strains.

Although it is possible to derive constitutive equations from micromechanics and homogenization methods [8, 9, 10], the phenomenological approach initiated by Lemaitre and Chaboche [11] will be used in this paper. For the sake of simplicity, we focus essentially on the loss of stiffness due to damage and neglect the other phenomena. More sophisticated models for brittle heterogeneous materials exist, e.g. [12, 13], but as far as size effect is concerned their properties are not fundamentally different from the model which will be discussed.

The material is described by two state variables: the elastic strain tensor ε and the damage denoted D. We assume that D is a scalar which represents randomly oriented microcracks. The Helmholtz free energy of the material is written:

$$\rho\psi = \tfrac{1}{2}(1 - D)\Lambda\varepsilon{:}\varepsilon, \tag{2}$$

where Λ is the initial stiffness matrix and ρ is the mass density. The constitutive equations are derived from thermodynamics of irreversible processes [12].

$$\sigma = \frac{\partial\rho\psi}{\partial\varepsilon} = (1 - D)\Lambda{:}\varepsilon. \tag{3}$$

The damage variable varies monotonically from 0 for the virgin (unloaded) material to 1 at failure. The rate of energy Y dissipated due to damage is also calculated:

$$Y = -\frac{\partial\rho\psi}{\partial D} = \tfrac{1}{2}\Lambda\varepsilon{:}\varepsilon. \tag{4}$$

The evolution of damage is governed by a loading surface denoted generically $f(\varepsilon, \Lambda, \varepsilon_D)$ in which ε_D is the threshold of damage initiation and the damage variable is calculated from a non-associated evolution law $\dot{D} = \mathscr{F}(\varepsilon)$.

In the present form, the damage variable is supposed to be zero at the initial state and is independent of the number of initial defects in the material. Thus $D = 0$ denotes a reference state at which the stiffness Λ is measured. We assume that the amount of initial defects is small, then the onset of damage is the only material characteristic which varies with the initial defect concentration and Λ is independent of the reference state.

3. Probabilistic description of damage initiation

The effect of random defects inside the material may be introduced as a random distribution of elastic properties or damage thresholds in a micromechanical model (see e.g. [14, 15]). Each computer simulation is a realization of a random process and statistics require an extensive number of runs (roughly proportional to the number of degrees of freedom of the model) in order to obtain a representative average response of the material to given loads. Considering the present state of the art in computer technology, this type of approach remains practically out of reach. In many applications it is certainly more convenient to assume a given probability density of a random event such as damage initiation and to derive the most probable threshold at which the event will occur.

In the present case, we are interested in an evaluation of the initial damage threshold ε_D. For concrete, Mazars [16] defined this threshold as the state at which tensile strains have reached the limit:

$$\tilde{\varepsilon} = \varepsilon_D$$

in which

$$\tilde{\varepsilon} = \sqrt{\sum_{i=1}^{3} \langle \varepsilon_i \rangle_+^2}, \quad \langle \varepsilon_i \rangle_+ = \frac{\varepsilon_i + |\varepsilon_i|}{2}. \tag{5}$$

$\tilde{\varepsilon}$ is called the equivalent strain and ε_i are the principal strains. The equivalent strain defines the accumulated tensile strain in the material. We assume that initial defects start to propagate in the form of voids and cracks when tensile strains have reached a threshold, thus damage initiation is a function of $\tilde{\varepsilon}$.

In a volume ΔV divided in elements of volume δV, the distribution of ε_D is random. Denote $P_d(\varepsilon, \delta V)$ the probability that in the volume δV the equivalent strain is higher than the threshold ε_D. The probability that damage will not be initiated in δV is

$$P_{nd}(\tilde{\varepsilon}, \delta V) = 1 - P_d(\tilde{\varepsilon}, \delta V). \tag{6}$$

If we neglect any spatial correlation on ε_D, the probability of 'non-damage' in the volume ΔV is

$$P_{nd}(\tilde{\varepsilon}, \Delta V) = [1 - P_d(\tilde{\varepsilon}, \delta V)]^r, \tag{7}$$

with $\Delta V = r\delta V$. For large values of the integer r, (7) becomes

$$P_{nd}(\tilde{\varepsilon}, \Delta V) = \exp[-rP_d(\tilde{\varepsilon}, \delta V)]. \tag{8}$$

We now introduce two requirements:

1. For infinitely large values of $\tilde{\varepsilon}$ the volume δV must be damaged i.e. $P_d(\tilde{\varepsilon}, \delta V) = 1$, and
2. For equivalent strains which are less than a constant, say ε_{D_0} a material property, the volume δV will not be damaged. A possible expression of the probability of damage is

$$P_d(\tilde{\varepsilon}, \delta V) = k(\tilde{\varepsilon} - \varepsilon_{D_0})^m \, \delta V. \tag{9}$$

This formula is inspired from Weibull studies [19, 20] on the probability of fracture of brittle materials. k and m are material properties, m is often called the Weibull parameter. The major difference is that in our approach Weibull's weakest link theory is used to describe the initiation of damage and not the ultimate strength of the material which is a function of some other phenomenon including stress redistribution.

We consider now a volume of material subjected to any arbitrary state of stress. The following is inspired from Jayatilaka studies on the fracture of brittle materials [21]. In each volume ΔV_i, the equivalent strain is denoted $\tilde{\varepsilon}_i$ and

$$P_{nd}(\tilde{\varepsilon}_i, \Delta V_i) = \exp(-kr(\tilde{\varepsilon}_i - \varepsilon_{D_0})^m \, \delta V_i). \tag{10}$$

The probability that the volume V will not become damaged is computed identically to (7) with $V = p\Delta V_i$

$$P_{nd}(V) = \prod_{i=1}^{p} \exp(-k(\tilde{\varepsilon}_i - \varepsilon_{D_0})^m \, \Delta V_i). \tag{11}$$

For infinitely small values of ΔV_i, (11) yields

$$P_{nd}(V) = \exp\left[-k \int_V (\tilde{\varepsilon} - \varepsilon_{D_0})^m \, \mathrm{d}V \right]. \tag{12}$$

The most probable value of the damage threshold ε_D in the specimen of volume V is computed from statistics

$$\varepsilon_D = \int_0^{\infty} P_{nd}(V) \, \mathrm{d}\tilde{\varepsilon}. \tag{13}$$

We may also note that the specimen is linear elastic thus

$$\int_V (\tilde{\varepsilon}_i - \varepsilon_{D_0})^m \, \mathrm{d}V = (\tilde{\varepsilon}_M)^m \int_V h^m(x) \, \mathrm{d}V, \tag{14}$$

where $\tilde{\varepsilon}_M$ is the maximum value of $\tilde{\varepsilon} - \varepsilon_{D_0}$ over the volume V and $h(x)$ is function of both the geometry and the loading conditions. Substitution of (14) and (12) into (13) yields

$$\varepsilon_D = \frac{\Gamma(1 + (\tfrac{1}{m}))}{\left(k \int_V h^m(x) \, \mathrm{d}V \right)^{1/m}} + \varepsilon_{D_0}, \tag{15}$$

164   *J. Mazars et al.*

where Γ is the classical function:

$$\Gamma(x) = \int_0^\infty e^{-t} t^{x-1} \, dt. \tag{16}$$

$\Gamma(1 + 1/m)$ may be regarded as a constant and we can write (15) as:

$$\varepsilon_D = \left(\frac{W_0}{\int_V h^m(x) \, dV} \right)^{1/m} + \varepsilon_{D_0}. \tag{17}$$

Three constants appear in this equation: W_0, m, and ε_{D_0}. We will assume that ε_{D_0} is very small compared to usual values of ε_D, thus $\varepsilon_{D_0} \approx 0$. W_0 and m will be obtained from experiments on specimens of different sizes.

Remarks

1. If the equivalent strain is homogeneous (17) reduces to

$$\varepsilon_D = (W_0/V)^{1/m} + \varepsilon_{D_0}. \tag{18}$$

    We see here the effect of the volume of the specimen on the damage initiation.
2. For structures of identical volume and geometry subjected to different boundary conditions, $h(x)$ is different. It reflects the effect of the loading conditions on the initiation of damage as its probability is a function of the state of stress.

4. Evolution of damage – nonlocal damage model

We need to describe now the evolution of damage once it has been initiated. Studies on localization instability have shown that a model in which damage is local, i.e. $D(x) = \mathscr{F}[\tilde{\varepsilon}(x)]$ yields physically unacceptable results as far as the description of failure of structures is concerned [3]. Analytical solutions show that the damage zone is restrained to a region of zero volume and failure occurs without dissipation of energy, a consequence of which is a strong mesh dependency in finite element calculations.

Among the possible remedies to this problem, a nonlocal continuum with local strain models was proposed [17, 18]. We will use in this paper this type of constitutive equation specialized to continuum damage – the so-called nonlocal damage model. The internal state variable D at point x is a function of the average equivalent strain over a neighbourhood centered at point x

$$D(x) = \mathscr{F}(\overline{\varepsilon(x)}), \tag{20}$$

$$\bar{\varepsilon}(x) = \frac{1}{V_r} \int_V \alpha(x - s)\tilde{\varepsilon}(s) \, dV,$$

where V is the volume of the structure and α is a Gaussian weight function. For two dimensional structures

$$\alpha(\boldsymbol{x} - \boldsymbol{s}) = \exp\left(-\left(\frac{2|\boldsymbol{x} - \boldsymbol{s}|}{l_c} \right)^2 \right) \tag{21}$$

and V_r is the representative volume such that

$$V_r(\boldsymbol{x}) = \int_V \alpha(\boldsymbol{x} - \boldsymbol{s})\,\mathrm{d}V. \tag{22}$$

The length l_c in (21) called characteristic length is proportional to the smallest width of the region in the medium into which damage can localize. From experiments $l_c \approx 3\,d_a$ in which d_a is the size of the largest heterogeneity in the material [6].

The nonlocal damage model is not merely an artifact introduced in numerical computations to prevent spurious strain and damage localization. From the homogenization of an interacting crack system sited in an elastic continuum, it was shown theoretically that damage should be defined as a nonlocal variable [10]. During the localization process, crack interaction produces energy redistribution in the continuum which may lead to an increase of damage near the localization zone, even though at the local level, the crack density remains constant. The weight function was also derived theoretically (its form is close to a Gaussian function) along with the value of the characteristic length which is consistent with the experiments [6].

We introduce the loading function for subsequent damage at any point $\boldsymbol{x}$ in the form

$$f\left[\overline{\varepsilon(\boldsymbol{x})}\right] = \overline{\varepsilon(\boldsymbol{x})} - \mathcal{K}(\boldsymbol{x}). \tag{19}$$

In the reference state $(D = 0)$, $\mathcal{K}(\boldsymbol{x}) = \varepsilon_D$, afterwards $\mathcal{K}(\boldsymbol{x})$ is equal to the maximum average equivalent strain ever reached at the considered point of the continuum in the past loading history. This loading surface is inspired from the St. Venant criterion and is relatively well suited to describe damage in concrete [12, 16]. Next, the condition for the evolution of damage is written

$$\text{if } f(\bar{\varepsilon}) < 0 \text{ or } f(\bar{\varepsilon}) = 0 \text{ and } \dot{f}(\bar{\varepsilon}) < 0 \text{ then } \dot{D} = 0,$$

$$\text{if } f(\bar{\varepsilon}) = 0 \text{ and } \dot{f}(\bar{\varepsilon}) = 0 \text{ then } \dot{D} > 0.$$

We use an integrated evolution law of damage [12, 17],

$$D = \alpha_t \mathscr{F}_t(\bar{\varepsilon}) + \alpha_c \mathscr{F}_c(\bar{\varepsilon}) \tag{23}$$

with

$$F_i(\bar{\varepsilon}) = 1 - \frac{(1 - A_i)\varepsilon_D}{(\bar{\varepsilon})} - \frac{A_i}{\exp(B_i(\bar{\varepsilon}) - \varepsilon_D)}, \quad (i = t, c). \tag{24}$$

We call σ_+ and σ_- $(\sigma = \sigma_+ + \sigma_-)$ the tensors in which appear respectively only the positive and negative principal stresses and ε_t, ε_c the corresponding strain tensors

$$\left.\begin{array}{l} \varepsilon_t = \Lambda^{-1} : \sigma_+ \\ \varepsilon_c = \Lambda^{-1} : \sigma_- \end{array}\right\}. \tag{25}$$

The weight α_t and α_c are defined by the following expressions

$$\left.\begin{array}{l} \alpha_t = \sum_1^3 H_i \dfrac{\varepsilon_{ti}(\varepsilon_{ti} + \varepsilon_{ci})}{\tilde{\varepsilon}^2} \\[2em] \alpha_c = \sum_1^3 H_i \dfrac{\varepsilon_{ci}(\varepsilon_{ti} + \varepsilon_{ci})}{\tilde{\varepsilon}^2} \end{array}\right\}. \tag{26}$$

$H_i = 1$ if $\varepsilon_i = \varepsilon_{ci} + \varepsilon_{ti} \geqslant 0$, otherwise $H_i = 0$. The parameters A_t, B_t, A_c, B_c in the evolution laws, are identified independently from compression tests on cylinders, and bending tests on beams. Two different kinematics of damage are used depending on the sign of the principal stresses, α_t and α_c are the coefficients defining the contribution of each type of damage for general loading. In uniaxial tension $\alpha_t = 1$, $\alpha_c = 0$, $D = D_t$ and vice-versa in compression.

5. Numerical implementation

This nonlocal continuous damage model has been implemented in the general purpose finite element code CESAR developed at the Laboratoire Central des Ponts et Chaussées in France [22]. Since stresses and strains remain local, the structure of the code need not be changed. Before the calculation begins on a specific structure, the value of the damage threshold is computed and a connectivity table is constructed for each integration point in which the Gauss points which pertain to its neighborhood and the corresponding values of the weight function are given. A simple secant stiffness algorithm is used with a convergence test on residual forces (for more detailed information see [24]). In this presentation, the analyses of two dimensional structures will be discussed.

(a) Necessity of a probabilistic damage initiation threshold

In this part our intention is to show that calculations with a fixed damage threshold cannot predict the strength and behaviour of different kinds of specimens made with the same concrete. For this, we consider three types of beam, a small one $(10 \times 10 \times 40\,\text{cm}^3)$ and a large one $(15 \times 22 \times 160\,\text{cm}^3)$ loaded in three and four point bending (Table 1). The play rules are: parameters are identified to describe the behavior of one type of the specimens (large four point bending beams), and used for the other types. The parameters obtained from this identification are: Elastic characteristics: $E = 3\ 10^4\,\text{MPa}$; $v = 0.2$, damage initiation threshold: $\varepsilon_D = 5\ 10^{-5}$, damage evolution laws parameters: $A_t = 0.8$; $B_t = 2\ 10^4$; $A_c = 1,4$; $B_c = 1850$, characteristic length: $l_c = 30\,\text{mm}$.

Table 1. Geometry and load conditions for the specimens subjected to three point and four point bending

Beam	Height (mm)	Width (mm)	Length (mm)	Span (mm)	Interval betw. loads (mm)
	220	150	1600	1400	800
	220	150	1600	1400	0
	100	100	400	300	100

distances are in mm.

Table 2. Comparisons of the maximum load, obtained from experiments and computations with and without a probabilistic damage initiation threshold

Beam	Experiments: Ultimate load (average)	Calculations Without probabilistic damage thresh.	With probabilistic damage thresh.
	15330	15310 (ϵd = 5.E-5)	15335 (ϵd = 5.08E-5)
	9600	8400 (ϵd = 5.E-5)	9315 (ϵd = 6.04E-5)
	11300	7320 (ϵd = 5.E-5)	10510 (ϵd = 8.01E-5)

loads are in N.

The results given by this set of parameters are shown in Table 2. The average ultimate strength obtained from experiments is correctly estimated for the large beam in four point bending, and underestimated by 12.5 percent for the same specimen in three point bending and underestimated by 35 percent for the small beam. It would be necessary to change the parameters in order to obtain a good prediction. We have changed only the damage threshold according to the probabilistic determination of damage initiation introduced before. Parameters necessary for this are adapted so that a correct prediction of the response of the same reference beam is obtained: $m = 6.5$ and $W_0 = 10^{-22}$ mm^3. The computations show a bias with experimental data of 3 percent for the large beams in three point bending and of 7 percent for the small beam. This result clearly points out the usefulness of the probabilistic approach.

(b) Numerical prediction of size effect

First, we have compared the prediction of the damage model with size effect results on notched beams subjected to three point bending. Experiments were carried out by Horvath and Persson [23] on notched beams of three different sizes. The geometry of the beams and reported material characteristics are given (Fig. 2a). Calculations were conducted on meshes made of triangular, six-noded elements with 724, 902 and 912 degrees of freedom respectively. The non-available material properties which are necessary for the damage model were determined by fitting the calculation with the data on the smallest beam. We obtained $l_c = 12$ mm, $E = 32300$ MPa, $v = 0.2$, $A_t = 1$, $B_t = 4200$, $A_c = 1.1$, $B_c = 1900$, $m = 5.8$ and $W_0 = 3.25 \ 10^{-24}$ mm^3 Figure 2a shows the comparison of the computations against the experiments. We notice:
1. that the agreement is quite satisfactory for the peak loads,
2. that a good fitting of the post peak response is also obtained. This last point shows that the present model predicts fairly well the evolution of damage. As far as the damage initiation is concerned we noted that the damage threshold was approximately constant (variations of less than 10 percent) which proves that the notch is the most prominent defect in the beams. Figure 2b shows the comparison of our results with Bažant's size effect formula (1). For the three beams, the logarithm of the nominal tensile stress at failure σ_N computed from beam theory is plotted versus the logarithm of the size of the specimen.

$$\sigma_N = 3P1/(d - a)^2, \quad (P \text{ is the ultimate load}). \tag{27}$$

As it was already concluded by Saouridis and Mazars [24], the size effect obtained from numerical calculations with the damage model is consistent with this simplified formula which was compared to experiments on numerous occasions, e.g. [3].

Saouridis and Mazars also investigated the prediction of this model against experiments on unnotched beams of different sizes [24]. They found out that the introduction of a probabilistic damage threshold was of primary importance. The location of the defect which initiates failure is not controlled in this case. The numerical predictions of the response of the three point bending beams tested by L'Hermite [1] without a variation of the damage threshold would dramatically underestimate the size effect on the nominal tensile stress at failure σ_N. If the specimen size is changed from 1 to 10, the tensile strength would be diminished by only 10 percent while experiments indicate a decrease of about 50 percent. Computations which include a probabilistic damage initiation threshold are shown in Fig. 3 and compare very well with experimental data. Again, the determination of the material constants was made so that a good fit is obtained for the smallest specimen ($E = 37700$ MPa, $v = 0.2$, $A_t = 0.8$, $B_t = 17000$, $A_c = 1.4$, $B_c = 1900$, $m = 6.4$, $W_0 = 3.5 \ 10^{-22}$ mm^3, and $l_c = 18$ mm).

It is interesting to note that the damage initiation threshold ε_D varies from $1.16 \ 10^{-4}$ for the smallest beam to $0.612 \ 10^{-4}$ for the largest beam. A large part of the size effect is due to the variation of ε_D in this case. However, failure still occurs by propagation of a damage zone concentrated in the mid-span cross-section of the beam and the global size effect observed in Fig. 3 is a combination of probabilistic and deterministic size effects, e.g. due to damage evolution, redistribution, etc. Overall, the predictions are close to experimental data but the present model predicts that failure is initiated and concentrated in the critical cross-section of the beam as determined by elastic analysis. In reality, failure may occur in a cross-section which

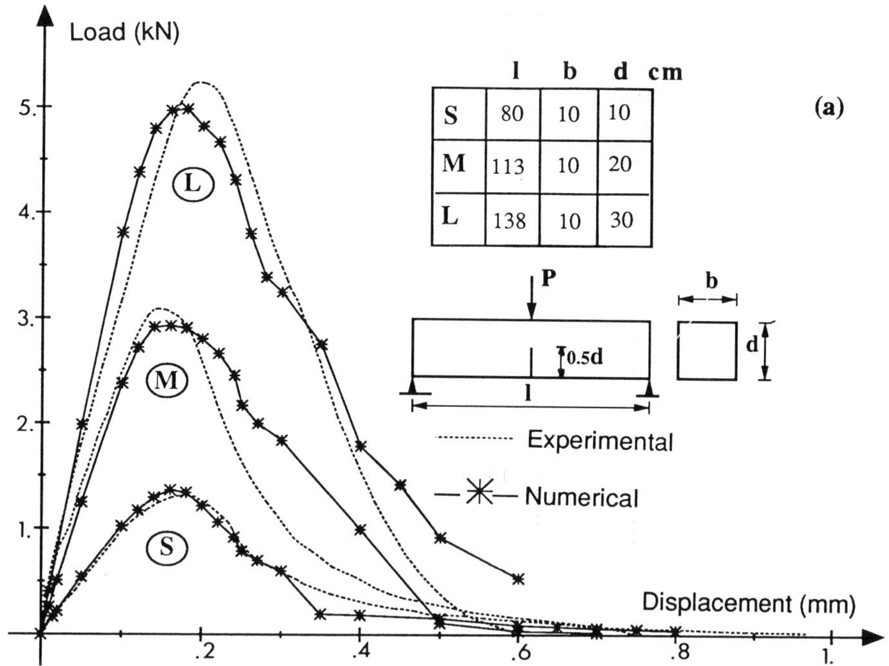

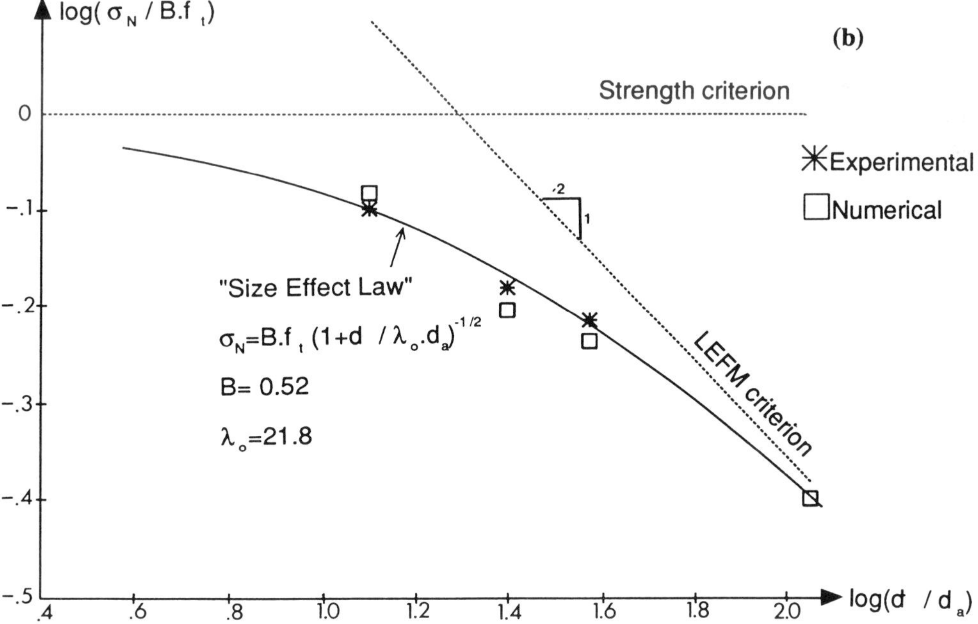

Fig. 2. Size effect on notched specimens; comparison with Horvath and Persson data [23]; (a) Force vs. displacements, (b) Tensile strength vs. size of the beam.

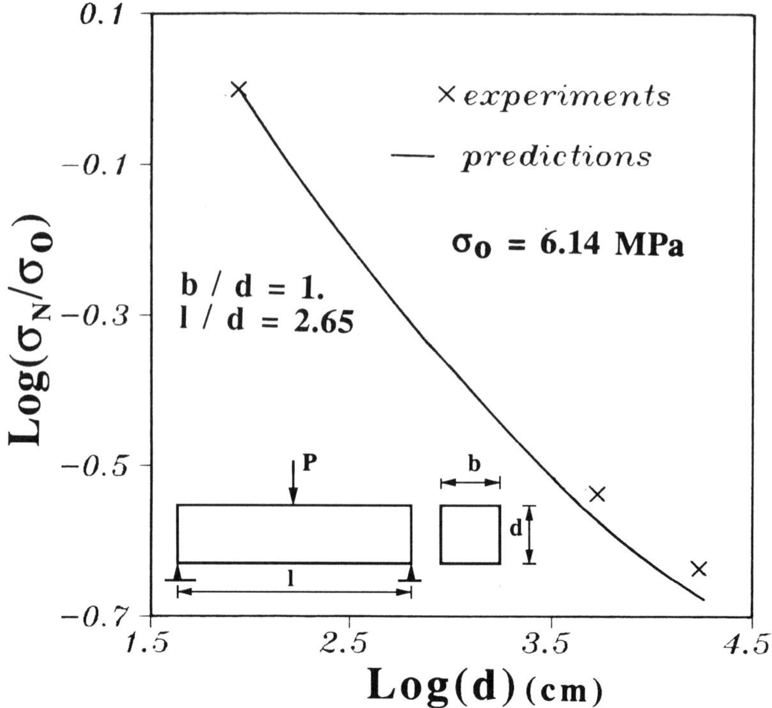

Fig. 3. Comparison with L'Hermite data [1] on unnotched beams: tensile strength vs. depth of the beam.

is not critical in the sense of beam theory because of the presence of a large defect on this cross-section. This cannot be modeled in the present formulation, only a complete probabilistic model would do so. Nevertheless, the probability that failure may occur in a cross-section which is not critical is included in our presentation. Namely, the function $h(x)$ in (15) provides this type of information and damage initiation depends also on the stress field applied to the structure. This effect is exemplified in Fig. 4a which shows the variation of damage initiation and nominal tensile stress at failure for a beam subjected to three point bending with a variable span. The results are non-dimensionalized to values obtained for a span of 280 cm. The beam has a square cross section $h = 70$ cm, the span varies from $l = 73.3$ cm to $l = 182$ cm and the length of the beam l' is constant $l' = 280$ cm. We note that the size effect on damage initiation is relatively strong. For the smallest span, failure occurs in compression due to stress concentration under the load, and tension on both sides of the beam. It explains why σ_N, *computed from a tensile strength criterion*, starts to increase, and then decrease as failure occurs only on the tensile side of the beam.

Finally, Fig. 4b shows the prediction of the size effect for specimens identical to L'Hermite's tests subjected to three point bending and whose scale ranges from 1 to 40. On this size effect plot we note that the plateau corresponding to limit analysis predictions on Fig. 1 has disappeared. This is due to the variation of damage threshold. Overall, for $\text{Log}(d) < 3.5$ the size effect is comparable (or even stronger) to LEFM predictions. For very large specimens, computations show that the tensile strength changes according to Weibull predictions with a slope of $1/m$ in a log/log plot. At this point failure under the load in compression due to stress concentration has also been initiated and the mode of failure of the beam is changing. Similar

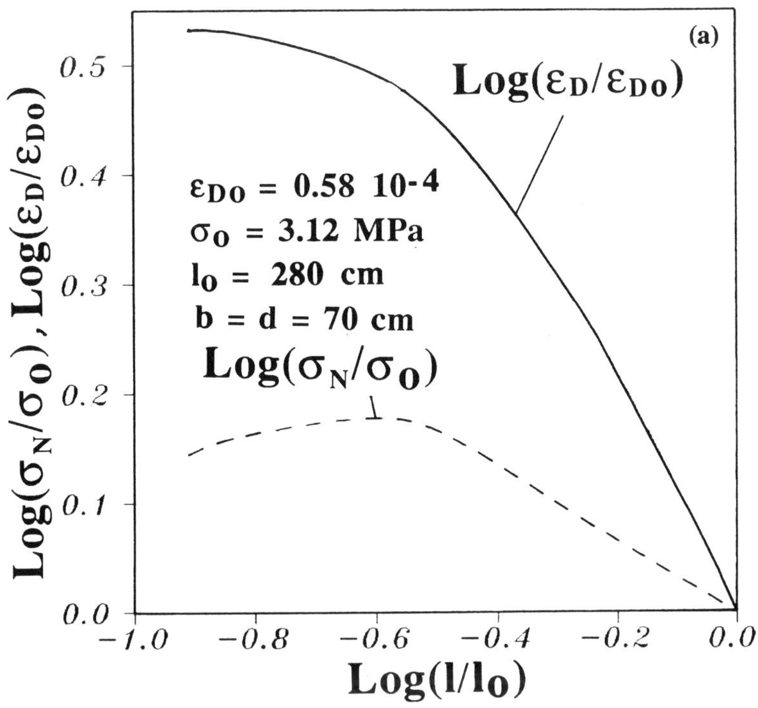

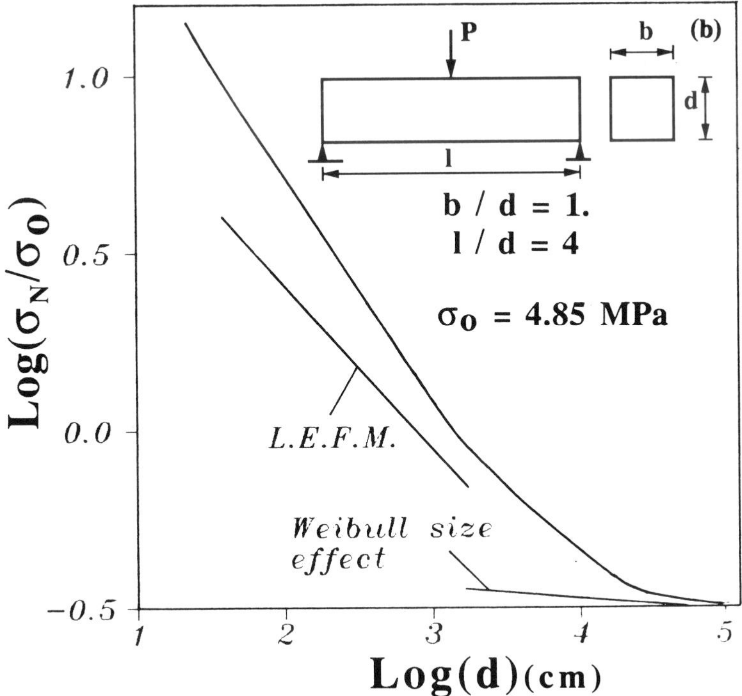

Fig. 4. Size effect on unnotched beams: (a) Effect of the span length on damage initiation and on the tensile strength; (b) Tensile strength vs depth of the beam.

effects, e.g. transition from failure in tension to failure in compression, have also been experienced on Brazilian tests conducted on specimens of large sizes. Without a probabilistic description of damage initiation the tensile strength should become almost constant for specimens of different sizes. However, one has to keep in mind that the failure process has changed, thus the estimate of the maximum strength from beam theory (27) is certainly erroneous as the maximum tensile stress criterion does not correspond to the real failure of the structure.

6. Conclusions

Two types of size effect have been distinguished in this paper:

1. A probabilistic size effect which affects damage initiation. Physically it can be related to the amount and location (with respect to the type of loading applied) of initial defects in a structure.
2. A deterministic size effect which is inherent to fracture propagation in brittle hetero-geneous materials. Because of heterogeneities, the fracture process zone lying ahead of a propagating crack in concrete is relatively large. It cannot be neglected as far as energy redistribution near the crack tip is concerned. The fracture properties of the material are not constant and depend on the ratio of the size of the crack versus the size of the structure.

These two aspects have been combined into a single model based on nonlocal continuous damage mechanics. Damage initiation is computed from the Weibull weakest link theory, and the evolution of damage due essentially to microcrack propagation in tension for concrete-like materials is described by a scalar, nonlocal, state variable which represents a loss of stiffness. Finite element computations and comparisons with experiments on notched and unnotched bending beams show that the model provides a realistic description of the variation of the maximum tensile strength of concrete when the volume, geometry, and boundary conditions of a structure are changed.

References

1. R. L'Hermite, *Annales de l'ITBTP* 309–310 (1973) 39–41.
2. V. Kadlecek and Z. Spetla, *Bulletin de la RILEM* 36 (1967) 175–184.
3. Z.P. Bǎžant, *Applied Mechanics Review* 5 (1985) 675–705.
4. Z.P. Bažant and M.T. Kazemi, *International Journal of Fracture* 44 (1990) 111–131.
5. Z.P. Bažant and M.T. Kazemi, Size effect in fracture of ceramics, Internal Report, Department of Civil Engineering, Northwestern University, Evanston, Illinois (1989).
6. Z.P. Bažant and G. Pijaudier-Cabot, *Journal of Engineering Mechanics ASCE* 115 (1989) 755–767.
7. J. Mazars and Y. Berthaud, *Comptes Rendus à l'Académie des Sciences* 308 série II (1989) 579–584.
8. D. Krajcinovic and D. Fanella, *Engineering Fracture Mechanics* 25 (1986) 585–596.
9. J.W. Ju, *International Journal of Solids and Structures* (1989) in press.
10. G. Pijaudier-Cabot and Y. Berthaud, *Comptes Rendus à l'Académie des Sciences* 310 série II (1990) 1577–1582.
11. J. Lemaitre and J.L. Chaboche, 'Mécanique des matériaux solides', Dunod-Bordas Ed (1985).
12. J. Mazars and G. Pijaudier-Cabot, *Journal of Engineering Mechanics ASCE* 115 (1989) 345–365.
13. S. Ramtani, Y. Berthaud and J. Mazars, in *Transactions of the 10th International Conference on Structural Mechanics in Reactor Technology*, A.H. Hadjian (ed.), AA SMIRT Pub, Vol. Q (1989) 9–14.

14. D. Breysse, *Journal of Engineering Mechanics ASCE* 116 (1990) 1489–1510.
15. P. Rossi and J.M. Piau, in *Cracking and Damage*, J. Mazars and Z.P. Bažant (eds.) Elsevier Publishers (1989) 91–103.
16. J. Mazars, *Engineering Fracture Mechanics* 25 (1986) 729–737.
17. G. Pijaudier-Cabot and Z.P. Bažant, *Journal of Engineering Mechanics ASCE* 113 (1987) 1512–1533.
18. Z.P. Bažant and G. Pijaudier-Cabot, in *Proceedings of the 4th International Conference on Numerical Methods in Fracture Mechanics*, A.R. Luxmoore et al. (eds) (1987) 411–432.
19. W. Weibull, *Journal of Applied Mechanics* 18, n°3 (1951) 293–287.
20. P. Kith and G. Diaz, *Res Mechanica* 24 (1988) 99–207.
21. A. Jayatilaka, *Fracture of Engineering Brittle Materials*, Applied Science Publishers Ltd, London (1979).
22. P. Humbert, *Bulletin de Liaison des Laboratoires des Ponts-et-Chaussées* 160 (1989) 112–115.
23. R. Horvath and T. Persson, 'The Influence of the Specimen Size on the Fracture Energy of Concrete', Report TVBM-5005, Lund Institute of Technology, Lund, Sweden (1984).
24. C. Saouridis and J. Mazars, 'Prediction of the Failure and Size Effect in Concrete via a Bi-Scale Damage Approach', *Engineering Computation* (in press 1991).

International Journal of Fracture **51**: 175–186, 1991.
Z.P. Bažant (ed.), Current Trends in Concrete Fracture Research.
© 1991 *Kluwer Academic Publishers. Printed in the Netherlands.*

Size-scale transition from ductile to brittle failure: structural response vs. crack growth resistance curve

ALBERTO CARPINTERI
Politecnico di Torino, Department of Structural Engineering, 10129 Torino, Italy

Received 11 June 1990; accepted 11 November 1990

Abstract. The concept of brittleness number is revised emphasizing the distinction between a stress-intensification treatment and an energy treatment. In the case of power-law hardening materials the relation between plastic stress-intensity factor and J-integral is translated into a relation between stress brittleness number and energy brittleness number. The structural response generally depends on both such numbers and, therefore, is not physically similar when varying the size-scale of the body. A connection is then established between structural response and crack growth resistance curve, the relevant parameters of a scale-invariant J-resistance curve being related to the energy brittleness number.

1. Introduction

The *structural response* is not physically similar when varying the size-scale of the body. A transition is experimentally evident from plastic collapse to brittle failure, passing through hardening, softening and snap-back behaviour. At the same time, the *crack behaviour* undergoes the same transition, from slow and stable growth to fast and un-stable propagation. Both structural response and crack behaviour depend also on the controlling parameter. If the feed-back quantity is the load, only global hardening behaviours may be detected; when the feed-back quantity is the loading point displacement, both hardening and softening behaviours may be controlled; when the feed-back quantity is the crack mouth opening displacement, (*CMOD*), even the snap-back or catastrophic branches can be captured.

In the present paper, a connection is established between *structural analysis* and the *crack growth resistance concept*. When the (*energy*) *brittleness number* s_E^* is lower than the intercept $\tilde{J}_0$ of the scale-invariant J-resistance curve, the structural response presents a sharp cusp catastrophe and the virtual J-resistance curve presents values always higher than the critical J_{IC}. On the other hand, when s_E^* is higher than $\tilde{J}_0 + \tilde{J}'(\xi - \xi_0)$, where $\tilde{J}'$ is the slope of the scale-invariant J-resistance curve, and ξ is the relative crack length ($\xi_0 =$ initial value), the structural response is very ductile and the J-resistance curve presents values always lower than the critical J_{IC}.

The brittleness number concept was proposed by the writer with reference to the stress-intensity factor for linear elastic materials [1–7] as well as for power-law hardening materials [8]. Subsequently, the concept was extended with reference to the strain energy release rate [9–15]. Herein it will be applied to the J-resistance curve. The same concept was proposed by the writer for reinforced concrete [16–19] and by Bažant for aggregate materials [20, 21].

It will be demonstrated that the brittleness number based on the stress-intensity factor is connected with the brittleness number based on the strain energy release rate in the case of linear elastic materials. Similarly, the brittleness number based on the plastic stress-intensity factor is connected with the brittleness number based on the *J*-integral in the case of power-law hardening materials. In general, the structural response depends on both these brittleness numbers, independent of the constitutive law of the material. An exception is provided in the case of the three point bending specimen of linear elastic material [12].

2. Scale effect on the structural response

2.1. Strain-hardening material

Let us consider a non-linear material with a double constitutive law in tension (Fig. 1). The former law (Fig. 1a) is a Ramberg-Osgood strain-hardening power-law:

$$\tilde{\varepsilon} = \tilde{\sigma}^n, \tag{1}$$

where $\tilde{\varepsilon}$ and $\tilde{\sigma}$ are non-dimensionalized with respect to the corresponding maximum values of strain and stress, ε_y and σ_y, and n is the Ramberg-Osgood exponent, $1 \leqslant n < \infty$. The latter law (Fig. 1b) is a strain-softening power-law of crack opening displacement, w, and stress, σ:

$$\tilde{w} = (1 - \tilde{\sigma})^{n'}, \tag{2}$$

where $\tilde{w}$ is non-dimensionalized with respect to the value w_c, beyond which the interaction between the crack surfaces vanishes, and n' is a softening exponent, $1 \leqslant n' < \infty$.

Let us assume that a solid body of the material described above is loaded by a concentrated force P, δ being the displacement of the force in its direction (Fig. 2). The functional relationship between force and displacement can appear as not fulfilling the monodromy condition, when snap-back instability is present (Fig. 3). On the other hand, such a relationship can always be

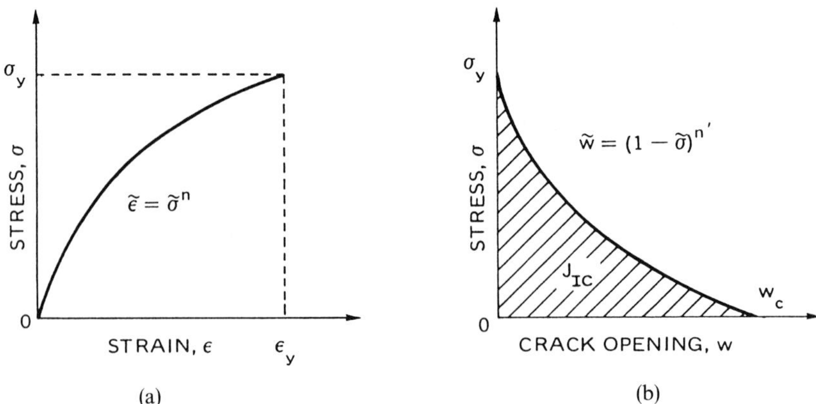

Fig. 1. Non-linear constitutive laws.

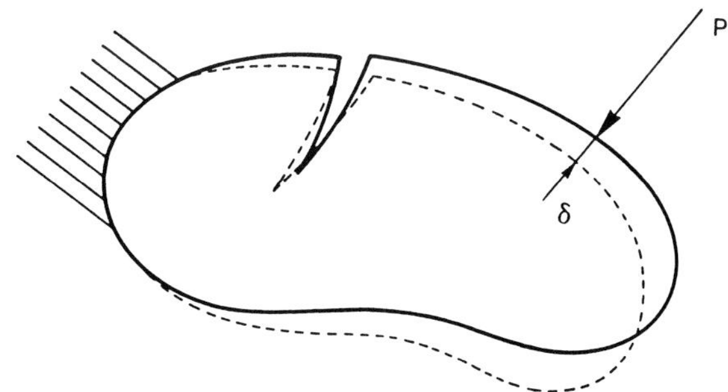

Fig. 2. Cracked body.

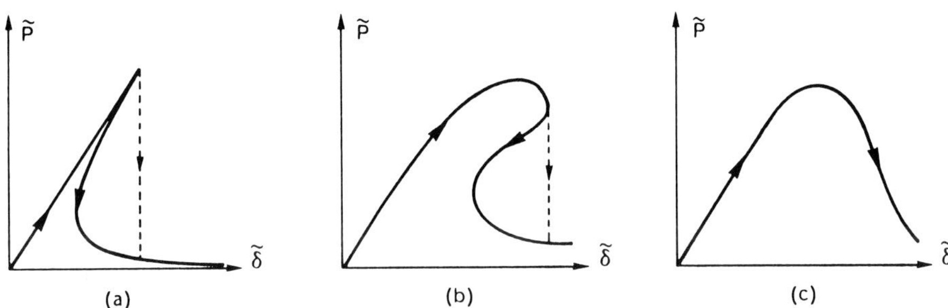

Fig. 3. Load vs. deflection paths: catastrophic (a), snap-back (b) and softening (c) behaviour.

put into the following general form [22]:

$$\pi\left(\frac{P}{\sigma_y b^2}, \frac{\delta}{b}; \varepsilon_y, \frac{J_{IC}}{\sigma_y b}, n, n'; \xi_i\right) = 0, \tag{3}$$

according to Buckingham's Theorem for physical similitude and scale modelling. In (3), b is a characteristic dimension of the body, J_{IC} is the area under the curve σ vs w (Fig. 1b), usually called the J-integral [23], ξ_i are the geometrical ratios necessary to describe the shape of the body.

If we fix the shape of the body and the shape of the non-linear constitutive diagrams in Fig. 1, the functional relationship (3) becomes

$$\pi(\widetilde{P}, \widetilde{\delta}; \varepsilon_y, s_E^*) = 0, \tag{4}$$

where $\widetilde{P}$ is the dimensionless load, $\widetilde{\delta}$ is the dimensionless displacement, and

$$s_E^* = \frac{J_{IC}}{\sigma_y b}, \tag{5}$$

is the (*energy*) *brittleness number* in the case of Ramberg-Osgood material.

The stress, strain and displacement fields at the crack tip present the following forms respectively [24, 25, 8]:

$$\sigma_{ij} = K_I^* r^{-1/(n+1)} \tilde{\sigma}_{ij}(\theta), \tag{6a}$$

$$\varepsilon_{ij} = \varepsilon_y \left(\frac{K_I^*}{\sigma_y}\right)^n r^{-n/(n+1)} \tilde{\varepsilon}_{ij}(\theta), \tag{6b}$$

$$u_i = \varepsilon_y \left(\frac{K_I^*}{\sigma_y}\right)^n r^{1/(n+1)} \tilde{u}_i(\theta), \tag{6c}$$

where K_I^* is the plastic stress-intensity factor, r is the radial coordinate, θ is the angular coordinate, and $\tilde{\sigma}_{ij}$, $\tilde{\varepsilon}_{ij}$, $\tilde{u}_i$ are non-dimensional functions of θ.

When $n \to \infty$ (rigid-plastic material) the stress-singularity tends to disappear, while the displacement tends to become only a function of θ.

It is possible to relate the plastic stress-intensity factor K_I^* to the J-integral

$$J_I = \int_{\mathscr{C}} W \, dy - \sigma_{ij} n_j u_{i,x} \, dl. \tag{7}$$

Namely, performing the integral along a circumference $\mathscr{C}$ centered in the crack tip, sufficiently small so that such a zone is dominated by the plastic singularity, yields [25, 8]

$$J_I = \left(\frac{K_I^*}{\sigma_y}\right)^{n+1} I_n \sigma_y \varepsilon_y, \tag{8}$$

where I_n is a well-known dimensionless function of n. The relation between J_I and K_I^* can be stated in the following form:

$$K_I^* = \sigma_y \left(\frac{J_I}{I_n \sigma_y \varepsilon_y}\right)^{1/(n+1)}. \tag{9}$$

If the (*stress*) *brittleness number* in the general case of Ramberg-Osgood material is defined as [8]

$$s^* = \frac{K_{IC}^*}{\sigma_y b^{1/(n+1)}}, \tag{10}$$

where K_{IC}^* is the critical value of the plastic stress-intensity factor K_I^*, from (5), (9) and (10) it is easy to obtain

$$s_E^* = I_n \varepsilon_y s^{*(n+1)}. \tag{11}$$

Equation (11) represents the relation between *energy brittleness number* and *stress brittleness number* in terms of the ultimate strain ε_y and the Ramberg-Osgood exponent n.

Equation (11) permits the transformation from (4) to

$$\bar{\pi}(\tilde{P}, \tilde{\delta}; s^*, s_E^*) = 0. \tag{12}$$

Equation (12) states that the $\tilde{P}$ vs $\tilde{\delta}$ path is the same if both the brittleness numbers s^* and s_E^* do not change. The reverse is not true.

2.2. Linear-elastic material

In the limit case $n = 1$, the Ramberg-Osgood material becomes linear-elastic (Fig. 4a), and the ratio

$$\sigma_y/\varepsilon_y = \sigma_u/\varepsilon_u = E, \tag{13}$$

represents the Young's modulus of the material.

On the other hand, for $n' = 1$, even the softening cohesive law is linear (Fig. 4b), with

$$J_{\text{IC}} = \mathcal{G}_{\text{IC}} = \mathcal{G}_F = \tfrac{1}{2}\sigma_u w_c. \tag{14}$$

If the ascending σ vs. ε law is linear elastic, the critical value of the J-integral must be equal to the critical value of the strain energy release rate, $\mathcal{G}_{\text{IC}}$, or to the fracture energy $\mathcal{G}_F$.

The functional relationship (3) between force and displacement (Figs. 2 and 3) in the case of linear-elastic material takes the particular form:

$$\pi\left(\frac{P}{\sigma_u b^2}, \frac{\delta}{b}; v, \varepsilon_u, \frac{\mathcal{G}_{\text{IC}}}{\sigma_u b}; \xi_i\right) = 0, \tag{15}$$

where v is the Poisson ratio. If we fix the shape of the body, (15) becomes

$$\pi(\tilde{P}, \tilde{\delta}; v, \varepsilon_u, s_E) = 0, \tag{16}$$

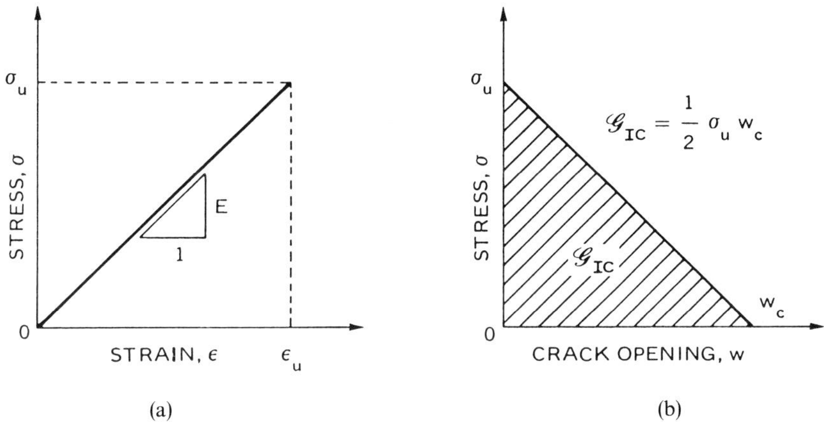

Fig. 4. Linear constitutive laws.

which corresponds to (4), if the (*energy*) *brittleness number*

$$s_E = \frac{\mathscr{G}_{IC}}{\sigma_u b},\qquad(17)$$

is introduced, as proposed by the writer [9–15].

When $n = 1$, $I_n = 1$, (9) is the generalized form of the well-known LEFM relation

$$K_I = \sqrt{\mathscr{G}_I E}.\qquad(18)$$

If the (*stress*) *brittleness number*, in the particular case of linear-elastic material, is defined as [1–7]

$$s = \frac{K_{IC}}{\sigma_u b^{1/2}},\qquad(19)$$

from (17), (18) and (19) it is easy to obtain [14]

$$s_E = \varepsilon_u s^2,\qquad(20)$$

which is the special case of (11).

If we take into account the relation (20) between s and s_E, through the ultimate strain ε_u, it is possible to present (16) in the following final form:

$$\bar{\pi}(\tilde{P}, \tilde{\delta}; s, s_E) = 0,\qquad(21)$$

with the influence of the Poisson ratio being negligible. It is the particular form of (12), for a material which exhibits linearity in the σ vs ε ascending stable stage.

2.3. Three point bending geometry

Based on the preceding sections, it is possible to state that in general, for linear or non-linear materials and arbitrary geometries, we have physical similitude and, therefore, the same $\tilde{P}$ vs. $\tilde{\delta}$ structural response only when both the brittleness numbers s^* and s_E^*, see (5) and (10), are unchanged.

On the other hand, in the particular case of linear material (Fig. 4) and three point bending geometry (Fig. 5), it was shown by the writer [13] that in order to have physical similitude, it is sufficient that the (*stress*) *brittleness number s*, see (19), remains unchanged. A numerical analysis based on the cohesive crack model and applied to the three point bending geometry is able to particularize (16) as follows [13]:

$$\pi\left(\tilde{P}, \tilde{\delta}; v, \frac{s_E}{\varepsilon_u}\right) = 0,\qquad(22)$$

and therefore, from (20)

$$\pi(\tilde{P}, \tilde{\delta}; s) = 0,\qquad(23)$$

the influence of the Poisson ratio again being negligible.

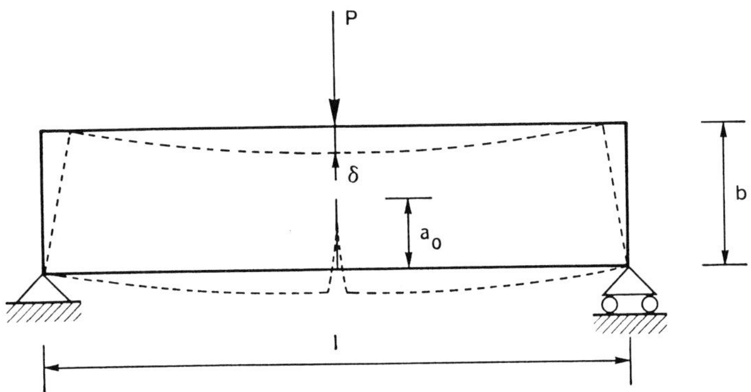

Fig. 5. Three point bending specimen.

If also the beam slenderness, $\lambda = l/b$, is involved in the analysis, (16) can take the form [13]

$$\pi\left(\tilde{P}, \tilde{\delta}; \nu, \frac{s_E}{\varepsilon_u \lambda}\right) = 0, \tag{24}$$

as is explained by a limit-analysis model proposed by the writer [11, 12], where the snap-back behaviour appears if

$$\frac{s_E}{\varepsilon_u \lambda} \lesssim \frac{1}{3}. \tag{25}$$

When s_E is varied over four orders of magnitude, from 2×10^{-5} to 2×10^{-2}, completely different structural responses are represented in Fig. 6a for an intially uncracked

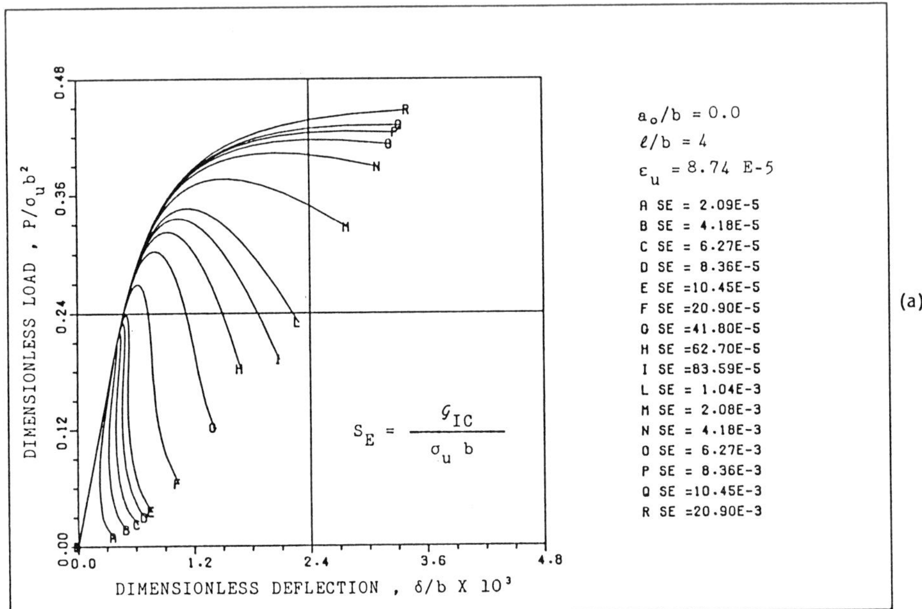

Fig. 6. Load vs. deflection diagrams by varying the energy brittleness number s_E.

182   *A. Carpinteri*

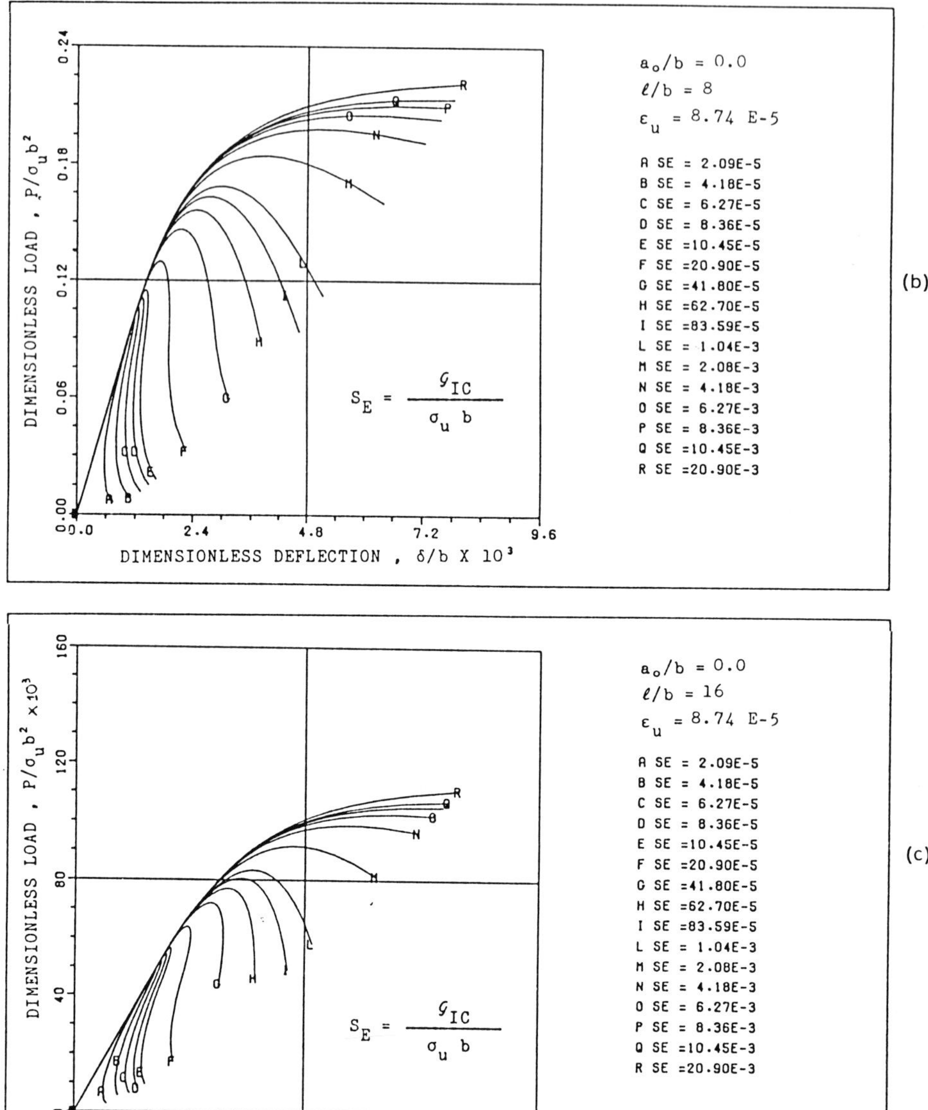

Fig. 6.—contd.

three point bending specimen, $a_0/b = 0.0$, of slenderness $\lambda = 4$, and an ultimate tensile strain $\varepsilon_u = 8.74 \times 10^{-5}$. The snap-back behaviour is predicted for $s_E \lesssim 10.45 \times 10^{-5}$, and then for

$$\frac{s_E}{\varepsilon_u \lambda} \lesssim \frac{1}{3.34}, \tag{26}$$

which is very close to the preceding approximate condition (25).

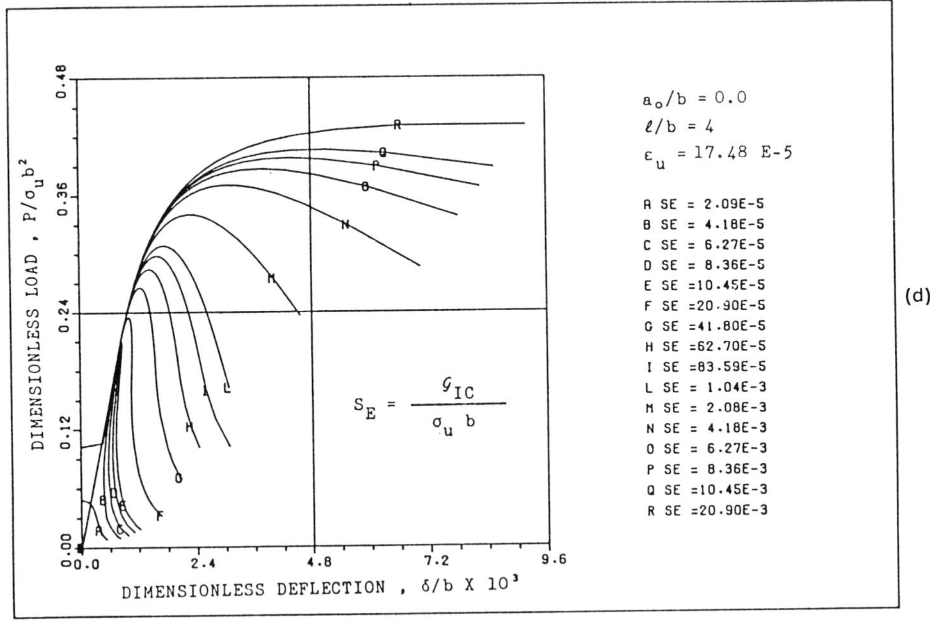

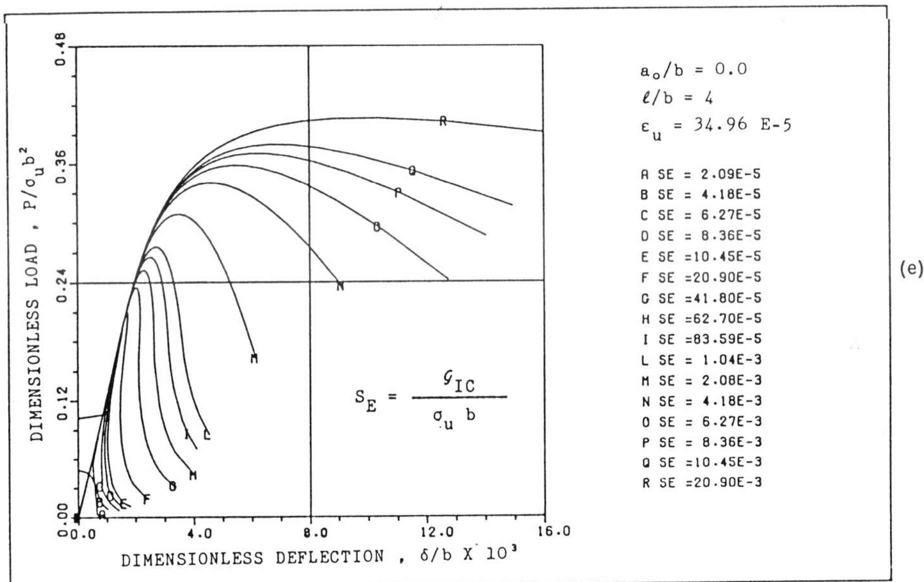

Fig. 6—contd.

The same variation of s_E is proposed in Fig. 6b, with $\lambda = 8$ and $\varepsilon_u = 8.74 \times 10^{-5}$. The snap-back behaviour is predicted for $s_E \lesssim 20.90 \times 10^{-5}$, and then (26) is still valid. In Fig. 6c, with $\lambda = 16$ and $\varepsilon_u = 8.74 \times 10^{-5}$, the snap-back behaviour is predicted for $s_E \lesssim 41.80 \times 10^{-5}$, and (26) is recovered once again.

In Figs. 6d and e, we have $\lambda = 4$ and ε_u respectively equal to 17.48×10^{-5} and 34.96×10^{-5}. In both cases condition (26) holds.

Recent experimental investigations on three point bending specimens of high strength concrete [26], brick [27] and rock [28] have confirmed the previous theoretical results. The extrapolation to mixed mode crack propagation in concrete was also considered [29, 30].

3. Scale effect on the J-resistance curve

The functional relationship between current value of the *J*-integral and relative crack length, is usually called the *J-resistance curve* and can be put into the following dimensionless form [22]

$$\Gamma\left(\frac{J_1}{\sigma_y b}, \frac{a}{b}; \varepsilon_y, n, n'; \xi_i\right) = 0, \tag{27}$$

where a is the current value of crack length. When J_1 reaches its critical value J_{IC}, the crack propagation becomes fast and uncontrollable.

Usually, at the beginning of the slow and stable crack growth, the J_1 vs. a relationship is an increasing function and can be considered as linear

$$\frac{J_1}{\sigma_y b} = \left(\frac{dJ_1/da}{\sigma_y}\right)\frac{a - a_0}{b} + \frac{J_0}{\sigma_y b}, \tag{28}$$

where $(dJ_1/|da)$ is the constant slope, J_0 the intercept on the J_1-axis and a_0 is the initial crack length. Equation (28) can obviously be rearranged into the form [22]

$$\tilde{J} = \tilde{J}'(\xi - \xi_0) + \tilde{J}_0, \tag{29}$$

where $\tilde{J}$ is the dimensionless *J*-integral and ξ is the relative crack length (Fig. 7). The constants $\tilde{J}'$ and $\tilde{J}_0$ are dimensionless and scale-independent. It follows that the slope of the J_1 vs. a diagram is constant varying the scale, i.e. $dJ_1/da = $ constant, and that the intercept J_0 is proportional to the size b. An experimental confirmation is found in [31].

On the basis of the previous assumptions, it is possible to state that above the size (Fig. 7)

$$b_{max} = \frac{J_{IC}}{\sigma_y \tilde{J}_0}, \tag{30}$$

stable crack growth ceases to occur and the brittle failure is achieved when the P vs δ curve is still in its linear course, whereas, below the size

$$b_{min} = \frac{J_{IC}}{\sigma_y[\tilde{J}_0 + \tilde{J}'(1 - \xi_0)]}, \tag{31}$$

unstable crack growth ceases to occur and the progressive slow crack growth develops up to the complete specimen separation. On the other hand, for $b_{min} < b < b_{max}$, stable (or slow) crack growth is followed by unstable (or fast) crack propagation. Even according to the *J*-resistance

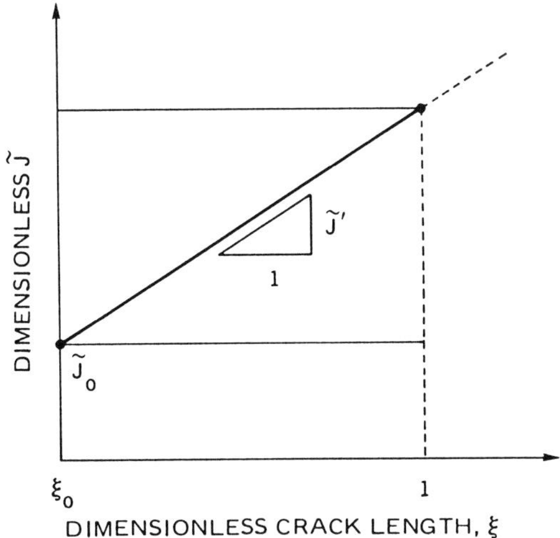

Fig. 7. Scale-invariant *J*-resistance curve.

approach, ductility or brittleness appears as a structural property rather than a material property.

Recalling the definition of (*energy*) *brittleness number* s_E^*, see (5), we can define a transitional interval between ductility and brittleness, following (30) and (31)

$$s_E^* \lesssim \tilde{J}_0 \quad \text{(brittleness)}, \tag{32a}$$

$$\tilde{J}_0 \lesssim s_E^* \lesssim [\tilde{J}_0 + \tilde{J}'(1 - \xi_0)] \quad \text{(transitional interval)}, \tag{32b}$$

$$s_E^* \gtrsim [\tilde{J}_0 + \tilde{J}'(\xi - \xi_0)] \quad \text{(ductility)}. \tag{32c}$$

Conditions (32) may correspond respectively to cusp-catastrophe, snap-back behaviour and strain-softening behaviour (Fig. 3).

4. Conclusions

1. A relationship between *stress brittleness number* (s^*) and *energy brittleness number* (s_E^*) is obtained (11) when both σ vs. ε and σ vs. w constitutive laws are non-linear. Each number is derivable from the other, once the ultimate strain ε_y and Ramberg-Osgood exponent n are known.
2. The structural response in the dimensionless plane $\tilde{P} - \tilde{\delta}$ depends on both the brittleness numbers s^* and s_E^*.
3. In the particular case of a linear elastic material and a three point bending geometry, the structural response depends only on the stress brittleness number s, if the geometrical ratios are fixed. Otherwise, it depends on the number $s_E/\varepsilon_u \lambda$, where λ is the beam slenderness.
4. The size-scale transition from ductile to brittle failure is captured also by the *J*-resistance approach. The relevant parameters of a scale-invariant *J*-resistance curve can be related to the energy brittleness number s_E^*, (32).

References

1. A. Carpinteri, in *Proceedings of the International Conference on Analytical and Experimental Fracture Mechanics*, Sijthoff and Noordhoff, Rome (1980) 785–797.
2. A. Carpinteri, *RILEM Materials and Structures* 14 (1981) 151–162.
3. A. Carpinteri, *Engineering Fracture Mechanics* 16 (1982) 467–481.
4. A. Carpinteri, *ASCE Journal of the Structural Division* 108 (1982) 833–848.
5. A. Carpinteri, C. Marega and A. Savadori, *Journal of Materials Science* 21 (1986) 4173–4178.
6. A. Carpinteri, *Engineering Fracture Mechanics* 26 (1987) 143–155.
7. A. Carpinteri, *Engineering Fracture Mechanics* 32 (1989) 265–278.
8. A. Carpinteri, *RILEM Materials and Structures* 16 (1983) 85–96.
9. A. Carpinteri, in NATO *Advanced Research Workshop on Applications of Fracture Mechanics to Cementitious Composites*, Martinus Nijhoff (1984) 287–316.
10. A. Carpinteri, *Journal of the Mechanics and Physics of Solids* 37 (1989) 567–582.
11. A. Carpinteri, *International Journal of Solids and Structures* 25 (1989) 407–429.
12. A. Carpinteri, *ASCE Journal of Engineering Mechanics* 115 (1989) 1375–1392.
13. A. Carpinteri and G. Colombo, *Computers and Structures* 31 (1989) 607–636.
14. A. Carpinteri, *International Journal for Numerical Methods in Engineering* 28 (1989) 1521–1537.
15. A. Carpinteri, *International Journal of Fracture* 44 (1990) 57–69.
16. A. Carpinteri, *ASCE Journal of Structural Engineering* 110 (1984) 544–558.
17. A. Carpinteri, *Magazine of Concrete Research* 40 (1988) 209–215.
18. C. Bosco, A. Carpinteri and P.G. Debernardi, *ASCE Journal of Structural Engineering* 116 (1990) 427–437.
19. C. Bosco, A. Carpinteri and P.G. Debernardi, *Engineering Fracture Mechanics* 35 (1990) 665–677.
20. Z.P. Bažant, *ASCE Journal of Engineering Mechanics* 110 (1984) 518–535.
21. Z.P. Bažant and P.A. Pfeiffer, *ACI Journal of Materials* 85 (1987) 463–480.
22. A. Carpinteri and G.C. Sih, *Theoretical and Applied Fracture Mechanics* 1 (1984) 145–159.
23. J.R. Rice, *Journal of Applied Mechanics* 35 (1968) 379–386.
24. J.R. Rice and G.F. Rosengren, *Journal of the Mechanics and Physics of Solids* 16 (1968) 1–12.
25. J.W. Hutchinson, *Journal of the Mechanics and Physics of Solids* 16 (1968) 13–31.
26. L. Biolzi, S. Cangiano, G. Tognon and A. Carpinteri, *RILEM Materials and Structures* 22 (1989) 429–436.
27. P. Bocca, A. Carpinteri and S. Valente, *RILEM Materials and Structures* 22 (1989) 364–373.
28. P. Bocca and A. Carpinteri, *Engineering Fracture Mechanics* 35 (1990) 241–250.
29. E. Ballatore, A. Carpinteri, G. Ferrara and G. Melchiorri, *Engineering Fracture Mechanics* 35 (1990) 145–157.
30. P. Bocca, A. Carpinteri and S. Valente, *Engineering Fracture Mechanics* 35 (1990) 159–170.
31. J.R. Gordon and R.L. Jones, *Fatigue and Fracture of Engineering Materials and Structures* 12 (1989) 295–308.